Culture in International Construction

Wilco Tijhuis
and
Richard Fellows

LONDON AND NEW YORK

First published 2012
by Spon Press

2 Park Square, Milton Park, Abingdon, Oxon OX14 4RN
711 Third Avenue, New York, NY 10017, USA

Routledge is an imprint of the Taylor & Francis Group, an informa business

First issued in paperback 2017

Copyright © 2012 Wilco Tijhuis and Richard Fellows

The right of Wilco Tijhuis and Richard Fellows to be identified as authors of this work has been asserted by them in accordance with sections 77 and 78 of the Copyright, Designs and Patents Act 1988.

All rights reserved. No part of this book may be reprinted or reproduced or utilised in any form or by any electronic, mechanical, or other means, now known or hereafter invented, including photocopying and recording, or in any information storage or retrieval system, without permission in writing from the publishers.

Notice:
Product or corporate names may be trademarks or registered trademarks, and are used only for identification and explanation without intent to infringe.

British Library Cataloguing in Publication Data
A catalogue record for this book is available from the British Library

Library of Congress Cataloging-in-Publication Data
Tijhuis, Wilco.
Culture in international construction / Wilco Tijhuis and Richard Fellows. – 1st ed.
p. cm.
Includes bibliographical references.
1. Construction industry. 2. Corporate culture. 3. International business enterprises. 4. Cultural relations—Case studies. I. Fellows, Richard, 1948– II. Title.
HD9715.A2T497 2011
338.8'8724–dc22
2011002342

ISBN13: 978-0-415-47275-3 (hbk)
ISBN13: 978-1-138-09289-1 (pbk)

Typeset in Sabon by Prepress Projects Ltd, Perth, UK

For our beloved families

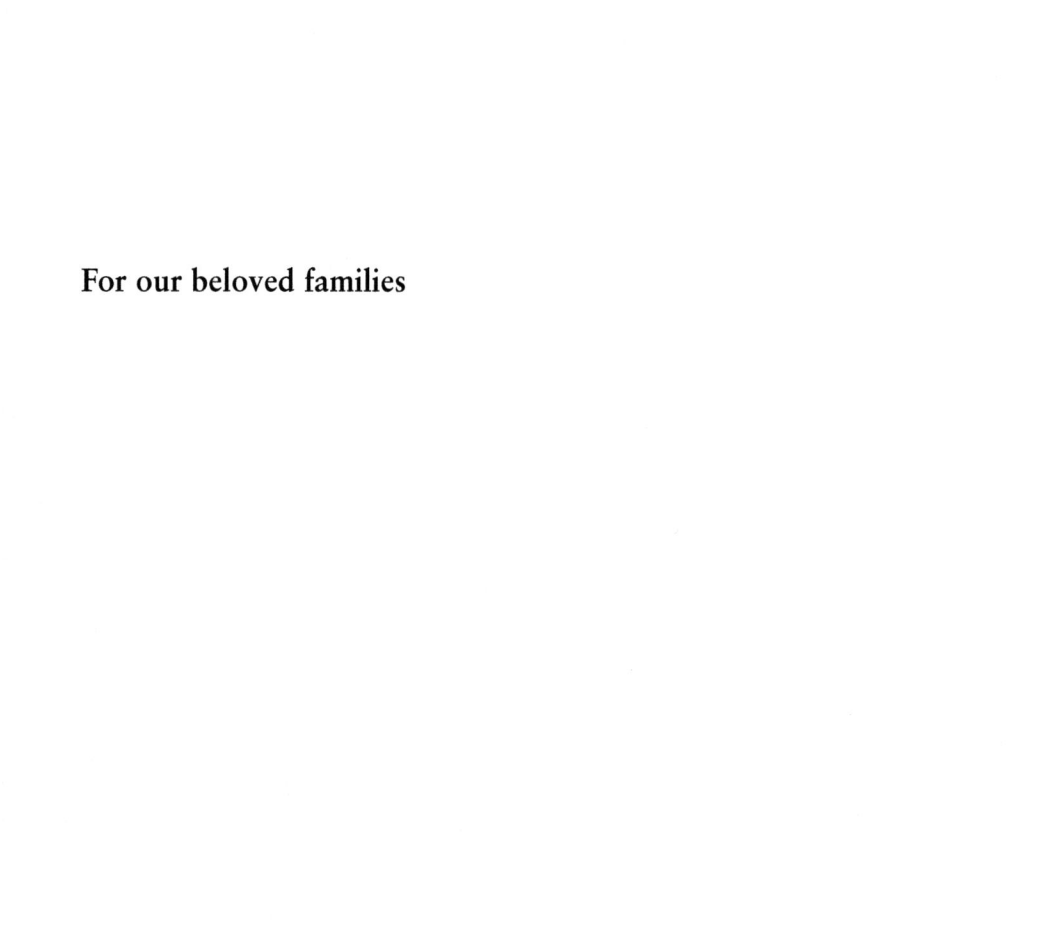

Men's natures are alike; it is their habits that carry them far apart.
Confucius
Chinese philosopher and reformer (551–479 BC)

Since we cannot know all that there is to be known about anything, we ought to know a little about everything.
Blaise Pascal
French mathematician and physicist (1623–1662)

Contents

List of illustrations   xi
Preface   xiii
Acknowledgements   xvii

1 Introduction   1
   Introduction 1
   Construction in an economy 4
   Construction, social institutions, social capital and human capital 8
   Culture 11
   Construction and culture 16
   Construction projects as joint ventures 18
   Overview of contents 20

2 Construction: a globalizing business   23
   Introduction 23
   Globalization 24
   Information technology (IT) and integration of markets 28
   Regional 'integration' 29
   Transparency and corruption 31
   Stakeholders, structures and processes 33

3 Culture's influences in construction: theory and applications   42
   Introduction 42
   Culture: definitions and nature 43
   Dimensions of cultures 48
   Differences and changes 57
   Construction culture 61
   Alliances 63
   Sustainability 68
   Conflict and disputes 73

4 The practice: international case studies — 79
Introduction 79
Case study 1: Developing a complex inner-city project 83
Case study 2: Construction of rationalized terraced housing 91
Case study 3: Subcontracting infrastructural and foundation works, a Polish case study 98
Case study 4: Tendering for developing a production factory 108
Case study 5: Designing a production factory 121
Case study 6: Organizing an international distribution structure for special building materials 137

5 Lessons learned — 151
Introduction 151
Lessons for sustaining contacts 151
Lessons for improving contracts 153
Lessons for preventing conflicts 154
Resumé 156

6 A future vision for culture in international construction — 157
Introduction 157
Developments in the construction industry 157
Developments in behaviour during critical incidents 167
Business culture's role in the construction industry 172
Resumé 175

Notes 176
References 177
Index 189

Illustrations

Figures

1.1	Culture spectrum	12
2.1	Holistic cycle of organizational development	34
2.2	The project realization process	35
3.1	Layers of culture	44
3.2	Inter-relations of (sub)cultures	46
3.3	Usual pattern of acculturation–adaptation to a new cultural context	47
3.4	Competing values and organizational cultures model	54
3.5	Denison's model of organizational culture (and leadership)	55
3.6	A conflict episode	76
3.7	Styles of conflict management and dispute resolution methods	77
4.1	The 3C-Model™ as a framework for investigating (behavioural) experiences in construction processes	81
4.2	Culture (Contact) and technology (Conflict) as drivers for changing construction processes (Contract) within project organizations, based on the 3C-Model™	81
4.3	Schematic situation of the project location	84
4.4	Groups of neighbours within this project, as parties involved in the process	86
4.5	Types of behaviour (positive and negative) of the parties involved in the three phases of the 3C-Model™	88
4.6	Example of a part of the housing project during its construction phase	92
4.7	Work phase in which additional waterproofing work is being done on one of the cellars	94
4.8	Types of behaviour (positive and negative) of the parties involved in the three phases of the 3C-Model™	96
4.9	Parts of the building foundations and parts of a cellar during construction	99
4.10	The whole building site in its rural location	101

xii *Illustrations*

4.11 Types of behaviour (positive and negative) of the parties
 involved in the three phases of the 3C-Model™ 106
4.12 The possible building site 109
4.13 An alternative building site 115
4.14 Types of behaviour (positive and negative) of the parties
 involved in the three phases of the 3C-Model™ 119
4.15 Investigating the proposed location 124
4.16 Types of behaviour (positive and negative) of the parties
 involved in the three phases of the 3C-Model™ 134
4.17 Ongoing building activities in Hong Kong, illustrating
 Hong Kong also as a city, considered to be one of the
 'economic gateways' to mainland China 140
4.18 The use of bamboo scaffolding is typical of the Hong
 Kong/Macau and Asian construction markets 141
4.19 Types of behaviour (positive and negative) of the parties
 involved in the three phases of the 3C-Model™ 147
6.1 Mechanisms influencing external motivation to use ICT in
 the construction industry 160
6.2 The existing building at Letisko International Airport in
 Bratislava, Slovakia 162
6.3 The building site for the new terminal building at Letisko
 International Airport in Bratislava, Slovakia, designed,
 engineered and realized by the efficient use of information
 technology tools 163
6.4 The completed new terminal building on Letisko
 International Airport in Bratislava, Slovakia 164
6.5 A relatively unusual sight in the construction industry: a
 laughing female construction professional on a construction site 167
6.6 The 'Thomas–Kilmann' model, showing the way a
 compromise in a possible conflict situation is reached or not 168
6.7 Schematic representation of social media networks
 as a tool for interconnecting people, thus stimulating
 influencing opinions and/or harmonizing human
 behaviour of (groups of) people 172

Tables

3.1 High-context/low-context (high-content) cultures 48
3.2 Perspectives regarding time 49

Preface

About the book

In the international world of the construction industry there is continuing dynamic movement, transforming (huge) amounts of capital, energy, materials and so on into magnificent results: houses, bridges, tunnels, offices, skyscrapers, highways, dams, dikes, pipelines and the like. In short, today's construction industry influences many parts of the economy and the environment, with all the pros and cons, regarding improved living circumstances, hygienic sanitary facilities, better road connections, reliable power supply, but also environmental influences, legal disputes, quality problems, cost and time overruns and so on. Nevertheless, the centre of this worldwide industry, with its local roots, is still the human being: clients, employees, managers, contractors, architects, engineers, users, investors and other stakeholders. Especially because all these individuals, and the organizations to which they belong, play their own roles, pursue their own interests, and do so within their own business-cultural backgrounds, this makes construction business a real people business. That also explains the central theme of this book, *culture in construction*, which considers the importance of business-cultural backgrounds and how business culture influences people and their organizations within the daily international construction business.

This book includes extensive insights from hands-on experiences in selected case studies, combined with a state-of-the-art overview of latest results from leading research, collected and conducted on a global scale.

About the readers

Because of its thorough description and mix of practical experiences and theoretical analyses of national and international aspects of culture's influences on modern construction processes, this book provides essential reading for professionals in the construction business (e.g. executives, managers, consultants, clients), for academics and for students. The subject 'culture in construction' is handled not just from the viewpoint of problems and threats, but especially pointing at opportunities and strengths in national

xiv *Preface*

and international business environments. Through its extensive list of useful references, it provides a valuable source for further study and improving awareness of this challenging topic.

About the authors

Dr.ir.ing. Wilco Tijhuis, BSc, MSc, PhD, has worked in the international construction industry since around 1987. After his bachelor's degree in construction technology, he began work in Germany and the Netherlands as a project coordinator for the contractor company Wessels Group. After his master's degree in construction management and development from Eindhoven University of Technology (the Netherlands) in 1992, he worked until 1997 as a project manager and business and market developer in the Berlin region for Kondor Wessels Group NV, an international contractor company at that time listed on the stock exchange (the present-day company Volker Wessels). Parallel to that professional period he also graduated as a PhD from Eindhoven University of Technology (the Netherlands). In 1997 he founded his own holding company, WT/Beheer BV (www.wtbeheer.com), and he is also a shareholder in NTGroep (www.ntgroep.com) and internationally active in concepts, processes and projects. Being a board member, he was and is regularly involved in development and construction projects in, for example, the Netherlands, Germany, the Middle East, Central Europe, South Africa and China. Since 1997 he has also been a part-time assistant professor at the University of Twente (www.utwente.nl), specializing in international construction processes, procurement and business culture. As a management professional and entrepreneur he regularly supervises post-doctoral MBA students at TSM-Business School (the Netherlands) in the field of construction and development processes, strategy and risk management. Since 2004 he has been a joint coordinator of the international research platform CIB W112 'Culture in Construction' (www.cibworld.nl). Wilco is a member of the Dutch Royal Institute of Engineers (KIVI). He has written several papers, reports, book chapters, journal articles and contributions to international conferences, and has lectured and presented to students, academics and professionals in the worldwide academic community as well as in the international construction industry.

Richard Fellows, BSc, PhD, FRICS, FCIOB, MCIArb, is Professor of Construction Business Management at Loughborough University, UK; previously, he was a Professor in the Department of Real Estate and Construction, The University of Hong Kong, and Professor of Culture in Construction at Glasgow Caledonian University, UK. He graduated from the University of Aston and has worked as a quantity surveyor for several major contractors. Richard has a PhD from the University of Reading, has taught at a number of universities in UK and other countries and was coordinator for research in construction management for the Engineering and Physical Sciences Research Council in the UK. His research interests concern economics, contracts and law, and management of people in construction – especially cultural issues as drivers of behaviour and performance. He was a founder and for many years was joint coordinator of the CIB group, W112 'Culture in Construction'. Richard has published widely in books, journals and international conferences and is qualified as a mediator.

July 2011
Rijssen (the Netherlands)
Loughborough (United Kingdom)

Acknowledgements

Working in the academic environment and in the international construction and development industry are challenging professions. As authors of this book, we both daily experience several actual challenges within both disciplines, as our readers will also do within their daily practices and backgrounds. This is often encouraged by working with colleagues and professional partners from several different cultural backgrounds, meeting them within international working activities and during travelling around the world. This makes every day still a challenge with continuously improving insights and experiences. The connection between these disciplines is in fact the central basis of this book:

> The construction industry itself and the professional people working within it.

Without this (dynamic) construction sector and the daily input of its enthusiastic professionals, there is hardly any reason for an applied science such as a construction management discipline to exist. On the other hand, the construction industry has the advantage of the availability of (academic) developments and output such as education and training programmes, technological and process innovations. In general, and especially within the context of writing this book, the authors therefore wish to thank all the parties involved for their willingness to open their archives for us, and for sharing their experiences with us.

Although originating from a broad scope of available case studies and experiences, the selected case studies in this book focus on critical moments during construction processes, in order to learn from them. It is worth noting that all the case studies described resulted in more or less satisfactory positive solutions for all the parties involved. The analysed (parts of the) construction processes, in which those solutions were realized, were for us as authors indeed a valuable source of experiences and lessons, seriously acting as 'food for thought'.

However, we do realize that lessons learned from these selective case studies are still just a small part of the quite complex world of (business) culture: *the differences and dynamics related to habits and daily behaviour of human beings within their (working) environment.* Therefore, we, as authors, want to emphasize the continuing challenge and need for further academic building and improving of theoretical models and structures, but preferably not without also collecting experiences from the daily practice of industry; thus, following the famous British saying that 'the proof of the pudding is in the eating'. We therefore wish that through this book we may give all our readers at least interesting ingredients for cooking their own pudding, which may possibly lead towards their own challenging recipe!

July 2011
Dr.ir.ing. Wilco Tijhuis (BSc, MSc, PhD)
Rijssen (the Netherlands)
Professor Richard Fellows (BSc, PhD, FRICS, FCIOB, MCIArb)
Loughborough (United Kingdom)

1 Introduction

Introduction

It is indisputable that construction constitutes a vital part of any country's economy. Perhaps more importantly, construction also constitutes an essential component of every society, no matter what its level of so-called development. Those two self-evident observations provide the rationale for this text; not really the former aspect, which has been addressed extensively over the years by many specialists, but primarily the latter, which, although not ignored, has received only scant, passing attention.

Of course, it would be foolhardy to suggest that the two aspects of economy and society are separate – they are inexorably intertwined, both with each other and with further aspects, notably technology, law and politics. It has long been acknowledged that construction is a labour-intensive industry, although measures of labour intensity, degrees of labour intensity and forms of labour intensity vary around the globe and over time. Not only are the physical artefacts output by the industry of great importance but so are the socio-economic processes involved in producing the infrastructure and buildings and in maintaining their stock in useful condition. What that suggests is that understanding the structure of the industry, and its interrelationships with other social and economic activities and institutions, is essential to appreciating construction's roles and contributions at all levels – macro, meso and micro – as well as in relation to those individuals involved in, and associated with, the industry.

Generally, construction provides 'producers' goods', with housing being the notable borderline subsector; however, for all outputs of the industry, the demand is 'derived' – the goods are demanded not for their own sake but for what they contribute to other activities of, or close to, direct consumption. That means that the contributions which construction outputs make to final demand activities are really what is demanded: shelter, security and so on. That nature of demand means that performance of the industry in (producing and) assembling the final outputs is judged against what is really being demanded by the immediate consumer (usually, the 'commissioning client' or the 'employer' in a construction contract). Further, the activities of the industry in maintaining the products in useful condition and their final disposal are becoming recognized to be of great importance to the effective and efficient functioning of society and to striving for sustainability.

In that context, it is important to appreciate the constitution of demand and demand transmission processes. Demand is human want, need or/and desire coupled with both the preparedness and the ability to pay the price – that combination renders demand effective in the world's capitalist markets. Demand may be expressed to potential and actual suppliers directly by final consumers (as in purchasing a cup of coffee in a café) or indirectly through a private sector intermediary (such as a developer developing an office block) or the public sector (as in public housing, schools and hospitals). Private sector intermediaries forecast demand for potential final outputs (such as office buildings) and then express demand to the construction industry based on the forecasts which they speculate will yield them the best return. The public sector intermediaries forecast need (e.g. for housing), subtract appropriate stock to yield the forecast requirement, review that requirement in the context of the finance available and express the resultant as demand (for public housing) to the construction industry. However, there is a notable blurring between the public and private sectors of economies through the array of privatization initiatives – notably, the various forms of public–private partnerships (PPPs).

The 'traditional' separation of design and construction (project assembly) processes, and the further separation of production of goods and components to be assembled, leads to a highly complex set of relationships on any project and, most especially, to interdependence between functionally separate organizations for achieving the final output. Examining only the site assembly arrangements significantly extends the number, type and functions of separate organizations operating on any project – today, in an increasing number of countries, all construction operations are let to subcontractors; 'main' contractors do little, if anything, more than manage the operations of subcontractors, thereby rendering all projects to be realized via management contracting, irrespective of the procurement process adopted formally.

Further, there is a common confusion of what performance is being judged. The process of realizing a construction project and the resulting product are distinct entities although, of course, inexorably related. The product output – the building, bridge, road or whatever – which is likely to have a useful life for a long time (50 or more years?) should be evaluated in terms of its fulfilling the requisite functions in use; that is real 'project performance'. The realization process (determining the brief, design and construction) should be judged not only in terms of its efficiency, as is most common (time, cost and quality – achieved compared with predicted), but also, and more importantly, in terms of its effectiveness in delivering the product.

It is well known, and widely acknowledged, that perceptions of performance can be highly variable; it depends on who is asked and when those persons are asked. But what about the 'hard' metrics of performance? Although widely assumed, and supposed, to be objective and accurate measures of particular aspects of performance, they are, in fact, 'negotiated realities' which reflect human components such as technical knowledge and expertise, negotiating skill and bargaining power; thus, they are only approximations to the 'objective realities' which they are believed to represent. Those, usually overlooked, considerations apply to project final accounts, durations and so on.

Almost all of the reports across the world which are the results of studies of the industry and investigations of its performance criticize the levels of performance which the industry achieves. Commonly, much of that alleged poor performance is attributed to fragmentation: the separation of functional activities. Conversely, from various perspectives, separation of functions is welcomed as 'division of labour' (Smith, 1789/1970; Taylor, 1947) or specialization – thereby fostering expertise and increases in productivity. However, construction is an increasingly complex activity, technically and managerially, so functional specialization seems inevitable (and potentially essential and advantageous). The critical element is 'integration'. Lawrence and Lorsch (1967) ably articulate the vital combination of specialization and integration to secure good performance of (complex) activities undertaken by diverse contributors. However, the emerging theory of complexity (e.g. Lucas, 2005, 2006) suggests that the performance issues may be an inevitable result of construction's being chaotically complex. Similarly, prospect theory (Kahneman and Tversky, 1979; Kahneman, 2003), through addressing incremental approaches to value – rather than the usual approach (based on Bernoulli, 1738/1954), which considers total value (utilities) – may be helpful in gaining understanding of dissatisfaction with performance.

The structure of any country's construction industry is dynamic through, necessarily, responding to workload, the mix of work types, technological changes and other environmental variations, notably legislation. Especially since the 1970s, the various forms of market capitalism have been in pronounced ascendancy, most obviously at the macro level with the restructuring of the former Soviet Union, Eastern Europe and, more recently and differently, retaining much overt and covert central control, the People's Republic of China and other countries in Asia. Throughout, and following, the Reagan/Thatcher era, the nature of capitalism in many countries has altered with depletions of previously public sector provisions and extensive privatization (what became recognized in Britain as the 'demolition of the welfare state'). For much construction, the trends are reflected in the ascendancy of procurement arrangements of PPPs, private finance initiative (PFI) and so on, the typology of concession contracting arrangements to reduce the capital contribution and involvement of the public sector.

At meso levels, a widespread change relates to the structure of the industry which reflects increases in specializations and changes in risk levels and distributions (enhanced by employment legislation). Amongst (design) consultancies, two diametrically opposite trends are occurring, new discipline practices are emerging, including interior design, landscape architecture, project management and programming, whilst, concurrently, multi-disciplinary practices are forming to endeavour to achieve more coordination and integration amongst the specialisms involved. For construction activities, the trend is stronger and uni-directional towards subcontracting whereas in facilities management outsourcing of the management, maintenance and related operational minor works on buildings-in-use is commonplace.

Today, construction (required) processes and procedures are rife with contradictions. 'Partnering' is advocated extensively and framework agreements are increasingly common whilst, concurrently, the alleged virtues of competition must be pursued to accord with legislative requirements. Many procurement procedures and contract forms are geared to traditional 'main contracting' structures while the reality of industry operational structuring is 'management contracting'. Common procedures assume that the architect (or engineer) carries out project management for the commissioning client, even though a separate project manager may be in post. Given these, and further, errors of organizational appreciation/awareness, it is hardly surprising that good coordination and high levels of collaboration are all too rare and that performance suffers in consequence.

That, naturally, leads back to the fundamental issues of performance – what is it, for whom, and so forth? A typology of performance, for any economic activity, not just for construction, comprises the three categories of generic business performance, technical performance and relational performance; it is only the technical performance which is activity-specific. In the current environment, private sector (and quasi-private sector, as in China) organizations stress growth and profit (really, profitability) as their primary performance criteria. Thus, technical and relational performance (or other performance metrics) are pursued in support of generic business performance. Indeed, such a hierarchy has a well-established theoretical and empirical basis: 'there is one and only one social responsibility of business . . . to increase its profits' (Friedman, 1970). Baumol's (1959) investigations yielded that, as management and ownership are separate in modern corporations, those organizations operate to maximize turnover (growth, the criterion of managers) subject to a minimum profit constraint (to satisfy the owners). Hutton (2002) discusses the power of fund managers of global institutional investors to impact the behaviour of public companies by requiring them to provide a non-decreasing stream of dividends to retain the fund managers' investments.

Construction in an economy

Apart from physiological survival requirements, construction is one of the oldest, and most basic, of human activities. The importance of construction surrounds everyone, often too evidently to be noticed. Commonly, the products of the industry constitute the greatest form of stock of wealth for most societies, organizations and individuals. Of course, the monetary valuations of built assets vary dependent upon prevailing economic circumstances but the real contribution of buildings and civil engineering products to human welfare (their real – use – value) is indisputable.

It is important to be aware of the nature of construction products, from the point of view of their functional purposes. What is often examined relates to the construction industry and its activities – construction supply (see below); but what is, in many ways, of much greater long-term importance to society is the stock of construction products (roads, railways, houses, factories, etc.). In both numerical terms and current valuation terms, stock dwarfs supply. Of course, the concepts of stock and supply are linked: the more enduring (long life, adaptable) the stocks, the lower the annual supply required (exacerbated in those developed countries which have declining populations). Functional change requirements of infrastructure and buildings tend to boost supply requirements (as do size increases and structural changes in populations), but in an era of increasing awareness of resource restrictions (hence, attention to 'greening' and 'sustainability'), attention to durability and easy adaptation assume greater importance.

The UK economy, by way of illustrating the place of construction in a 'developed' economy (in current, Western terms), is quite typical in demonstrating the importance of construction. In 2007, the output of the construction industry contributed over 8 per cent of gross domestic product (GDP) at market prices and almost 54 per cent of gross domestic fixed capital formation (GDFCF), which itself contributed over 16 per cent of GDP. The output was mainly produced by over 192,000 private sector firms, which provided employment for almost 1.3 million people, of whom 750,000 were operatives. Insolvencies of construction companies and partnerships constituted over 14 per cent of the total (12 per cent in 1996) but bankruptcies of individuals, self-employed persons and the like in construction had experienced a steady decline to only 3.5 per cent of the total (over 12 per cent in 1996) (Office for National Statistics, 2009a,b).

Of the (approximately) 2.2 million persons (HSE, 2009a) who worked in the construction industry in Britain during 2007/08, 72 lost their lives (HSE, 2009b) and 3,764 incurred a major injury (HSE, 2009c). 'In the last 25 years, over 2,800 people have died from injuries they received as a result of construction work' (HSE, 2009a).

> The latest survey of self-reported work-related illness (SWI) carried out in 2007/08 estimated that 88,000 people whose current or most recent job in the last year was in construction suffered from an illness which was caused or made worse by this job.
>
> (HSE, 2009d)

The statistics record only the reported injuries and so on, although some estimates are included; however, there remains concern over non-reporting of injuries (which may be quite significant) and may be very extensive in some countries. 'Workplace injuries and work-related illness accounted for an estimated 1.0 and 1.7 million working days lost (full-day equivalent) respectively in 2007/08, with corresponding rates of 0.45 and 0.77 days per worker' (HSE, 2009e); approximate extrapolation from those statistics indicates that the direct cost to the industry is about £1,000 million; adding indirect costs would increase that estimate considerably!

The mix of output, by type of work, in 2007 was about 58 per cent new work and 42 per cent repair and maintenance. Of the new work, only just over 10 per cent was infrastructure, 34.5 per cent was housing (82.5 per cent private sector), 21.5 per cent private sector commercial and 7 per cent private industrial; public sector housing and non-housing constituted almost 21 per cent of total new work. Of the total repair and maintenance work (housing and non-housing), 67 per cent was carried out by the private sector (Office for National Statistics, 2009a).

Within the total context of 'economic development' patterns exhibited by countries around the world, the general pattern is for infrastructure to be developed first (demonstrating its role as a primary factor enabling other developments to take place effectively and efficiently), followed by a dominance of construction of new buildings, and a final phase of dominance of refurbishment, rehabilitation and repair and maintenance activities. National statistics do not always fully reflect the work distributions and so on, because of the definitions used and the methods and accuracy issues in the collection of data. Commonly, much refurbishment and rehabilitation work is categorized as 'new work'; the contributions of 'self-build' and 'do-it-yourself', as well as the (sometimes very large) activities of the 'informal sector' (the 'black economy') are not or cannot be included.

Hillebrandt (2000), for example, discusses an important aspect of macroeconomic management, as practised by many governments for much of the last century: the importance of the public sector as a source of construction work – historically, around 50 per cent in total, but varying amongst construction sectors (for example infrastructure and housing). However, following the increase in various privatizations, the proportion of construction work carried out for the public sector in many countries has declined – in the UK, in 2007, to about 32 per cent of all work. Through the multiplier mechanism in the economy, given the contribution of construction to the total economy and given the government's importance for construction demand, by either expanding its demand to stimulate or shrinking its demand to suppress (and funded through fiscal means), governments used the construction industry to help regulate the economy in striving for growth and reasonable stability. Usually, monetary policy was used in tandem to reinforce the policy objectives. The policy instruments either stimulated or shrank employment, through the employment multiplier, as well.

However, in pursuing such post-war, Keynesian measures, which were largely abandoned in the 1980s with the widespread adoption of monetarism, governments had to take notice of the accelerator principle. The accelerator operates on capital (producers') goods for which there is derived demand and where the stock of goods is large relative to annual supply. It assumes that the relationship between the level of supply of the final output (the consumer good or service) and the amount of capital good required per unit of supply of the final output is, approximately, constant. The accelerator principle is that, if the supply of the final output increases (decreases), then the supply of the capital good required for its production will increase (decrease) by a greater proportion and earlier. The time lag depends on the time required to produce the capital good (realize the building). The accelerator is examined per period of time and is subject to changes, in proportionate quantities and lead times, due to changes in technologies, production methods and so forth. As suppliers amend their capacities by various means (new production layouts, new technologies, outsourcing, expansion of total production facilities, etc.) in response to forecasts of demand for their final output, their confidence in those forecasts is critical in determining their actions.

So, it is not surprising that, with the extents of lead times necessary for the realizations of construction projects and the variable and reducing durations of economic business cycles, the timings of policy initiatives have been, on occasions, significantly out of step with actual levels of economic activity, resulting in periods of pronounced 'stop–go', exacerbated rather than, as intended, balanced by the economic policy instruments exercised.

In many countries, the years since 1980 have witnessed extensive amounts of privatization (transfers of activities from the public sector to the private sector) and in diverse forms. In the UK, provision of public housing was virtually curtailed following a moratorium on construction of new public housing ordered by the then Secretary of State for the Environment, Michael Heseltine. However, as many public housing projects were under construction and others at advanced stages of design or letting for construction, the full effects on the construction industry occurred over some time as those projects were completed – hence making it easier for the industry to cope. In the twenty-first century, privatization continues through the numerous PPP processes, including the PFI in the UK. The extensive uses of such concession contracting arrangements have put further restructuring pressures on the construction industry.

In the last quarter of the twentieth century, redistributions of risks relating to construction projects, changes in the economic climate, increasing speed of change, amendments to taxation systems, and legislation relating to enhancing security of employment for employees all contributed to the restructuring of the construction industry. Increases in costs of labour, technological advances and requirements for shorter construction periods helped promote the change towards total subcontracting of construction activities. In many countries, effectively, all projects are constructed through management contracting – 'main contractors' do not carry out construction operations themselves but manage subcontractors to do so; in consequence, employment by large firms (previously, traditional, main contractors) has declined and changed structurally (from operative labour to administrative, technical and professional [APT] employees).

The nature of financing projects under concession arrangements has made fundamental changes to contractors' operations: instead of the contractors receiving funds at short intervals following the progress of the works (by the interim progress payments terms), they incur enormous levels of expenditure in constructing each project for which they receive revenue over a predetermined, long period during which they operate the project (or similar). Further, the pragmatic necessities of forming consortia, particularly for mega projects, has moved contractors towards being assemblers of financial providers. (For further discussions see *Engineering, Construction and Architectural Management*, 1997; Gruneberg and Hughes, 2006.)

The changes in risk distributions and industry structure have also yielded redistributions of power as well as making significant alterations to operating processes. Most estimating concerns soliciting, selecting and assembling bids for work, and materials supply packages from subcontractors and suppliers – but, with that change, there is a redistribution of power regarding which main contractor is likely to be awarded the project (under competitive tendering) (for a more detailed discussion see, for example, Uher, 1990). Bidding for projects which include both major design components and the need to assemble financial packages to fund project realization is both a time-consuming and an expensive process; hence, only large organizations have the resources required. Also, given the vital role of success rate in the costs of tendering, and, hence, impact on organizational profitability, it is common for the number of bidders to be small (commonly three or fewer).

Thus, there are many formal, standard construction documents and procedures which are significantly out of step with realties and practices current in the industry (following continuums of structural changes).

Construction, social institutions, social capital and human capital

Construction is a social process and its outputs are social products; both products and processes continually impact on people's lives through providing shelter, employment, visual pleasure or intrusion, resource depletion, pollution and so on. The built environment is far from a modern phenomenon; it has existed, although not in its present forms(!), for many thousands, if not millions, of years. The products of the industry are readily identified and analysed as articles (stock items) of physical capital and of economic capital. However, the processes involved in realizing construction projects aid the development of social capital and provide accommodation for social institutions.

Social institutions constitute systems of social positions which are interconnected by social relations which enable the systems, through the actions of the members, to perform social roles. Thus, social institutions comprise norms of behaviour and limits of acceptability as well as consequences (perhaps, ostracizing or excluding for lack of manifest trustworthiness) for transgressors (those who are 'caught' – requiring policing and effective deterrent sanctions). Social institutions and their manifestations are dynamic, which challenges the common human preference for the status quo. Thus, the concept of social institutions embraces the theory of organizations and, so, includes both formal and informal organizations as well as the more nebulous institutions embodied in societies through manifestations of their cultures. The social institutions may be formalized (as in a professional institution) or highly informal (as in the norms of behaviour in a society or culture).

Social institutions constitute important contextual elements within which human activities occur. Especially for construction management, as a field of application of social sciences in a particular technological context, the prevailing nature of society, and the interactions of societies through internationalization and globalization, impact on what issues arise, their importance, and whether and how they are addressed – as well as by what methods and with what results. In many locations, the impacts of political control are significant, whether local, national or 'federal' (e.g. USA, EU). Bachmann (2001), for example, discusses the role of social institutions in the generation of trust and of trustworthy behaviour, in conjunction with but beyond legal imperatives, in business in different countries; Scott (2001) provides a more extensive analysis of social institutions and their impacts.

An appreciation of the context provided by social institutions may be gained by examination of the profile of the national culture of a country (see, for example, Hofstede, 2009). Although such profiles may be quite rudimentary, of questionable content and, possibly, somewhat out of date because of dynamic evolution, they nonetheless do constitute helpful indicators of how a society operates and what the generic values are through examinations of the manifestations of culture – notably, (norms of) behaviour, symbols (including buildings), language, practices and heroes. More specific appreciation may be gained from examining the cultures and climates within particular organizations, while bearing in mind that those are embedded within national cultures.

Internationalization and globalization bring further complexities with increasing movements of people across national borders and organizations operating in various ways in many countries, thereby not only generating much more diverse mixes of cultures but also enhancing the degrees and speeds of changes in cultures – hence, the concept of dynamic equilibrium is apposite in studying culture. We tend to obtain profiles of cultures from cross-sectional samples, thereby yielding 'snap-shots' of those cultures; as with any cross-sectional profile of a phenomenon which is temporally dynamic, the validity of the profile decays in direct relationship to the speed of change of the phenomenon. Thus, as cultures change increasingly rapidly, profiles which are several years old are likely to be rather unreliable.

10 *Introduction*

A further aspect of globalization by many organizations is how they endeavour to establish themselves globally (more usually, internationally in a large number of countries). Given the diversities amongst societies and natures of demands exercised, globalization requires diversity of operations and outputs to accommodate the differences. Such aspects of globalization have led to the concept of 'glocalization' (a globalizing organization and overall identity but tailoring itself to the requirements of each locality in which it operates), which relates to outputs (e.g. Coca-Cola) and to processes (e.g. construction contractors operate differently in differing locations: labour- or capital-intensive construction methods).

'Social capital is the sum of the resources, actual or virtual, that accrue to an individual or group by virtue of possessing a durable network of more or less institutionalized relationships of mutual acquaintance and recognition' (Bordieu and Wacquant, 1992: 119). The networks may arise through formal processes and therefore may be documented (as in associations between organizations); alternatively, they may be informal and arise from contacts of individuals as well as through the formal organizations. Such networks add to the social capital of organizations and/or individuals through fostering cooperation and are effective and efficient channels for information flows which are highly valuable in transmissions of business intelligence. Organizations use the network structures as a resource to assist governance of their relationships. Further, gaps within and between networks may constitute areas of opportunity which may be exploited through establishing links with additional organizations.

The concept of social capital and networks of relationships is particularly apposite in construction given its human-intensive nature, the transience of projects (product realizations), and the fragmented structure of interdependent specialist functionary organizations. However, the social capital and associated networks not only constitute an information and cooperation resource for those involved but also operate to forge norms of behaviour. Those norms act as constraints to comply with the behavioural expectation of others in the network and so foster higher levels of investments (more risk taking) because the behavioural norms constrain potential opportunistic behaviour.

The more fully connected, 'closed', a network is, the more and closer the links between members (organizations) and so the greater their inter-reliance and social capital. Thus, the more a construction organization gets involved with a project, the more closed the network with others on that project becomes. However, perhaps one of the underlying problems for construction projects is that the formal relationships, as dictated by standard procedures and forms of contract, discourage the formation of extensive relational networks but channel information flows through particular actors, thereby constraining the forming of networks. Pragmatically, such difficulties are partially overcome by the 'informal systems' (Tavistock Institute of Human Relations, 1966) as the usual channels through which communications occur and with the behavioural norms with which participants usually comply to realize the projects.

For the industry generally, following Bachmann's (2001) discussion, the norms and constraints provided by the social institutions yield limitations on behaviour within that relatively open network. For construction, especially larger and highly specialist organizations which are likely to come into contact and relationships with each other frequently on projects, and so operate under oligopolistic competition, their recognition of interdependence is well documented, as in the behavioural discussion of the operation of the 'hypothesis of qualified joint profit maximisation' (Lipsey, 1989: 243), in which the 'tacit agreements' over behaviour correspond to the norms in networks of social capital.

Social capital may be analysed using three dimensions: structural, relational and cognitive (Nahapiet and Ghoshal, 1997). The structural dimension concerns the positions of actors in the structure of social relationships; the relational dimension concerns the features of the relationships – notably trust and trustworthiness; the cognitive dimension comprises norms of common understanding of terms and norms of behaviour – and so relates to culture very closely.

Human capital comprises the skills and abilities of persons, either individually or aggregated in organizations. Thus, certain elements of an individual's stock of human capital are due to genetics through that person's inherent abilities and traits (e.g. perfect pitch) whereas other elements are acquired or extended through education (understanding the economics of bidding) and training (producing a critical path programme for a project).

Thus, human capital relates to individuals' abilities, social capital concerns relationships between individuals, and physical capital comprises artefacts; the common thread is that all forms of capital are desirable on account of their actual and potential contributions to the production of wealth – things which have value because they are wanted by persons.

Culture

As international activities grow and globalization extends, there is an increasing awareness that similarities and differences in preferences, behaviour and human choices are vitally important. If people are different in appearance, it is common to expect them to be different in behaviour; conversely, if people appear similar, they are expected to behave similarly. Those usual expectations are reflections of culture – ethnic and national. However, cultures occur at other levels too – notably, organizational.

Within societies, even if of the same ethnic group and socio-economic stratum, differences may be quite stark and important. For instance, engineers work by synthesis, and prefer to learn in a similar manner, whereas architects tend to be reductionist – they determine the total scheme and then consider progressively more detailed aspects ('deconstruction'), which,

12 Introduction

similarly, is reflected in their learning preferences – so it is quite a challenge to teach a group comprising both engineers and architects effectively and efficiently! Hence, in such an apparently culturally similar group – the class – there could be three levels of culture interacting: national, discipline and student. In an industry working situation – such as on a construction project – where members of various disciplines are assembled to design and construct a project, the significant variations in their preferred learning and working styles may hamper collaboration and communication of ideas and proposals. Naturally, such difficulties are compounded geometrically when mixes of people from different ethnic origins and/or international operations are involved.

Thus, within any cultural typology and taxonomy, there are likely to be multi-level (sub)categorizations; even a group which, at first, appears structurally simple and coherent may comprise considerable cultural diversity and complexity. From a theoretical stance, which has significant implications in practice, especially for effecting change, there is often considerable confusion among 'culture', 'climate' and 'behavioural modification' (see Figure 1.1). So the primary, vital message is 'don't be fooled by appearances'.

Additionally, there is a notable tendency to consider systems and processes as simple and linear when, in reality, they are highly complex and non-linear (commonly, design of buildings is acknowledged, and advocated, to be iterative).

By definition, culture is a construct which can exist only for collectives of people and can be described, loosely, as 'how we do things around here'

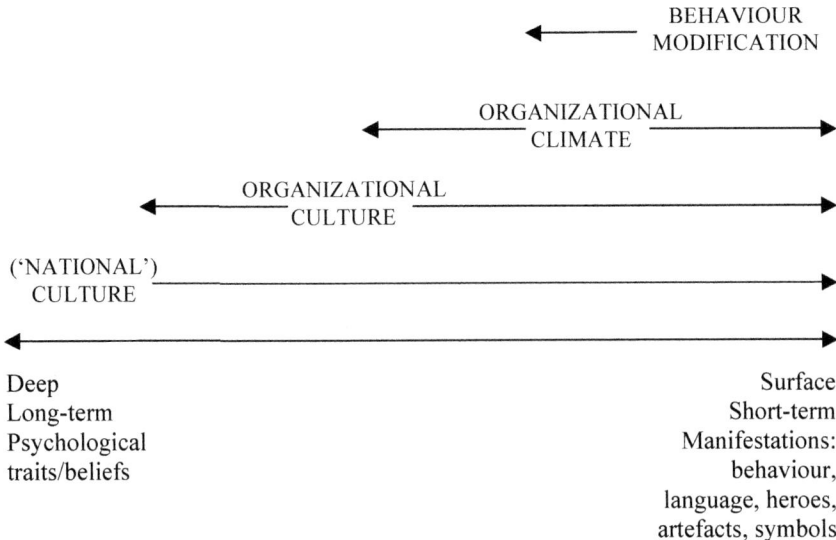

Figure 1.1 Culture spectrum (Fellows, 2006a). Note: boundaries between cultures, climates and behavioural modifications are fuzzy.

(Schneider, 2000). However, much more is involved: culture is not only *how* things are done; the construct of culture is much more extensive and includes *what* is done, *why* things are done, *when*, *by whom* . . . Despite those shortcomings, the description does have a behavioural focus and so draws attention to that primary manifestation of culture.

Kroeber and Kluckhohn (1952) define culture as:

> patterns, explicit and implicit, of and for human behaviour acquired and transmitted by symbols, constituting the distinctive achievements of human groups, including their embodiment in artefacts; the essential core of culture consists of traditional (i.e. historically derived and selected) ideas and, especially, their attached values; culture systems may, on the one hand, be considered as products of action, on the other as conditioning elements of future action.

Cultures are dynamic (Hatch, 1993) and change over time, usually quite gradually and predictably but subject to occasional perturbations. The prevailing culture not only impacts on the development of persons within it but is also reshaped by those persons. Thus, culture seems to be a primary manifestation of social constructivism in that people within, and so constituting, a culture behave according to the norms which they perceive to operate in that culture, moderated by their own personalities. The norms of behaviour within a culture extend over a range, the size of which depends upon the tolerance and diversity within that culture; also, because of dynamism, the range of norms shifts over time and is likely to alter in width too.

Culture provides context for making meaning, setting goals, taking action, constructing images and forming identities. In construction, a vital consideration is the impact of culture on what performance is achieved and measured against predetermined, culturally bound, targets, perhaps determined through value management or engineering exercises early in the project realization phase; those performance targets become the expectations against which performance is judged – notably by clients. Thus, for both design and construction activities, application of the concept of glocalization is essential for success through ensuring sensitivity to and accommodation of local requirements; impositions of alien designs and production methods will not work, as found by Toyota in employing Japanese production methods in the USA – effective and efficient in Japan but enormously less so in the USA (see Womack, Jones and Roos, 1990). Adaptations to allow for and incorporate cultural differences are essential.

A particular difficulty with culture is determination of the boundary. In systems theory, determination of four attributes of the system boundary is critical:

- location – where the boundary is; distinguishing what is in the system and what is outside it;

14 *Introduction*

- permeability – how easy it is to move across the boundary; varying from an open system with almost no detectable boundary to a closed system in which the boundary is almost impenetrable;
- flexibility – how easy it is to change the location of the boundary; often involving changing the size and scope of the system;
- effects – consequences (changes) which arise from crossing the boundary.

Primarily, a system is determined from its 'primary task': the function which the system is to fulfil. Pragmatically, the operation of any system occurs within constraints or parameters: things which have restricting impacts on the operating of the system. Together, those variables determine the system contents and, hence, the location of the boundary of the system: what is within the system and what is outside it. Essentially, a system is a transformation process which converts inputs, drawn from the environment, into outputs which are returned to the environment. As effectiveness and efficiency are required, control through feedback or feedforward mechanisms is essential in further relating the system to its environment (regulation).

A further consideration in determining culture (boundary) is derived from the concept of coherence and so can be articulated in terms of delineation of groups. Drawing on statistical concepts, groups are delineated through metrics of aspects of similarity which are critical to the issue(s) under study: a group contains items which are similar, relative to other items, hence within-group (intragroup) variance is small compared with between-group (intergroup) variance.[1] That assumption of internal coherence is of widespread importance in conceptualizations of cultures, especially national cultures, but, on account of migration, globalization and so on, is increasingly questionable. Thus, Au (2000) asserts that variations along dimensions of culture within national (or organizational) boundaries may be quite large, and even greater than variations along those dimensions between populations. From a social or cultural perspective, many boundaries are quite arbitrary and developed through historical 'power politics'; further, many boundaries are subject to quite sudden and major changes (e.g. the splitting of the USSR or corporate amalgamations and take-overs). Project organizations are, by definition, transient.

Culture, at whatever level, is recognized as a complex phenomenon thanks, largely, to its being 'measured' through its manifestations whilst being grounded in deep-seated human beliefs (assumptions) and values; further aspects of complexity are added by its concerning collectives of people and its dynamism. It is rather surprising, therefore, that, to date, studies of culture have not employed the emerging field of complexity theory. 'Complexity theory states that critically interacting components self-organise to form potentially evolving structures exhibiting a hierarchy of emergent system properties' (Lucas, 2006). Complexity theory combines systems theory, organic thinking and connectionism and so incorporates system variability, dynamism and non-reductionism such that the constantly evolving whole system, including parts which are in dynamic equilibrium, is synergistic (the whole is greater than the arithmetic sum of its parts) (see, for example, Lucas, 2005).

Lucas (2005) notes that complex systems are composed of independent agents who have autonomy; the systems are organic and therefore metabolize to produce their own parts and maintain them, which also renders them innovatively adaptive to the environment (autopoietic). The systems are defined by the connections between their parts, rather than by the parts themselves. Thus, leadership and control emerge through self-organization within the system. The function of the system is fuzzy at the outset but emerges as the system operates and, through interaction with the environment, values and semantics are constructed through probabilistic matching of the system with its environment. Consequently, in determining solutions, there are likely to be numerous and significant compromises determined by use of a plurality of techniques. Such overall fuzziness and the requirement to adopt a holistic perspective seems to yield a highly appropriate approach for studying construction projects – in particular, their cultural aspects and consequences.

The concept of cultural diversity has spawned the construct of cultural distance (see, for example, Kogut and Singh, 1988), which uses dimensions of culture, usually national, to constitute metrics by which distances are 'measured' and, then, aggregated to yield an index. Such indexes ignore the variability within individual cultures articulated by Au (2000), as well as requiring assumptions about commonality and compatibilities of the scales of measurement, and can, therefore, produce results which render expectations which are very different from a particular cultural encounter. Thus, it seems more apposite to examine cultural grouping in the way that scientists do when testing the safety of medicines: by assembling test groups which reflect the diversity in society. If pursued in 'cultural distance' studies, this could produce a readily usable three-point analysis quite easily (as for β-distributions in risk management; Fellows, 1996). Alternatively, it is likely to be much more revealing, and therefore useful, to consider measurements which have been made along each cultural dimension individually to yield a 'cultural profile indicator'.

Culture is a deep-seated social construct which can be examined only through scrutiny and 'measurement' of its surface manifestations; thus, commonly, it is overlaid with other constructs which make it difficult to determine what is culture and what is due to more superficial, but nonetheless important, socio-psychological factors. Those constructs, and their effects, get confused – as in so-called 'partnering' in construction.

Under the influences of internationalization and globalization, the debate regarding trends of societies along converging or diverging paths is important. Whereas certain manifestations exhibit convergence (shopping malls, hotels, airport terminals), others show enduring differences (religious buildings, languages, food – except 'fusion food'); there is 'no clear trend

16 *Introduction*

to convergence or divergence' (Boyer, 1996: 30). Evidence remains hugely mixed; some buildings are firmly grounded in local traditions and beliefs (Petronas towers) whereas others are 'locationless' (many of Mies van der Rohe's glass-clad office towers); English is the language spoken most widely (and increasingly so), although Chinese is spoken by the greatest number of people; but many communities strive to preserve and extend fluency in their particular local language (Gaelic, Welsh, individual Chinese tribal languages); similar convergence and divergence 'trends' are widespread in clothing, dances and dietary preferences.

The construction industry is involved with the realization, preservation and adaptation of important symbols of culture: buildings and infrastructure. Because considerable preservation of past/historic symbols takes place, such operations confirm the places of those symbols within the culture of current society (in the valuing of such artefacts from the past).

Figure 1.1 identifies and relates the main categories of constructs pertaining to cultures and their manifestations. Although the boundaries of the concepts remain somewhat fuzzy, it is important to appreciate the differences as well as the relationships between them. Fundamental to that understanding is awareness of the nature of culture (and, thence, requirements for changes in culture) as deep-seated in the human psyche and so requiring amendment of fundamental beliefs and values to bring about change – which, therefore, can occur only in the long term. Thus, culture can relate only to a social group (an institution) which endures in the long term.

For construction, and similar industries, failure to appreciate the nature of culture has spawned notions of cultural formation and change which are superficial and therefore can be effected in the short term. Hence, there has arisen a concept of 'project culture' which has been examined even for short-duration projects. Although it is appropriate to consider that long-term mega projects are likely to develop their own cultures, that is hardly the case for smaller projects of shorter durations. On such projects, the concept of 'project culture' is, essentially, an amalgam of the cultures of the constituent organizations (and societies) as transmitted to the project by their more powerful agents (managers, etc.) and perhaps more appropriately, therefore, could be referred to as 'project atmosphere'.

Construction and culture

There is an old joke, 'What is the difference between construction and live yoghurt? – There's culture in live yoghurt!'[2] However, given that culture is more than pursuit of certain forms of entertainment, behaviour, speech and the like, the fallacy of that joke is obvious; on the contrary, there is enormously rich culture and cultural diversity in the construction industry.

An organization's culture is grounded in the culture of its host (national) society – normally, the country/society of origin or of the primary, 'head office' location. Given international activities of construction firms, and globalization more generally, organizations are tending to become more multi-cultural and adopt glocalization. More ubiquitous communications and faster, cheaper and accessible transport extend such developments, including increasingly mobile and transient labour forces in all strata and roles in the industry. Of course, that is not only an international phenomenon; within the borders of many countries significant cultural differences exist, quite often between proximate locations.

Of course, what constitutes the construction industry varies from country to country – if only by examination of industry definitions used in national statistics. The concept of an industry, analogous to the statistical conception of a group, is a set of reasonably coherent, complementary production or service activities as judged from the perspectives of outputs and processes. Usually, the definitions take into account historical evolutions (such as the inclusion of open-cast coal mining in past definitions of the construction industry in Britain). However, a number of issues remain in determining the boundary of the industry, including whether design consultants are included, how to account for 'self-construction' (do-it-yourself, DIY), how to account for the 'unofficial economic activities' and so on. Within the industry, there are concerns over what data are collected and what statistics are produced, such as the inclusion of refurbishment and rehabilitation work as part of 'new construction work', the separate assessment of infrastructure (major projects can produce significant impacts – distortions – on statistics and trends), and the methods employed to account for the very many small firms and their activities.

Most industries in most countries (probably all) comprise a wide diversity of organizations by type, size and activity. Commonly, economists ask why that is so and how they survive successfully. However, given that they do, it is highly suggestive that there is also a diversity of organizational cultures, both vertically and horizontally, within any industry. Those organizational cultures, although all grounded in the (local version of) the national culture, also reflect the beliefs and value hierarchies of the influential members of the organizations, both past and current, to give rise to the differing organizational objectives and practices.

The dominant, common elements of the organizational cultures of construction project participants then become reflected in norms and practices (including contract forms and other regulatory instruments) relating to the industry and its activities (as for national societies).

Reviews of literature pertaining to culture and construction yield a picture of some concern. The theoretical bases, and the empirical studies and results which they underpin, are not widely understood and so are not used with rigour in application to the industry. That gives rise to misconceptions about what culture is and the dynamics of its operation at many levels – such as the notion of 'project culture' (see above). Additionally, the industry is given a vast array of cultural descriptors, many of which denote negativity, including 'macho', 'opportunistic', 'short-term', 'conflict/disputes' and 'claims-oriented'. Conversely, a number of initiatives have attempted to foster positive attributes and to combat negative aspects, including quality, safety, collaboration ('partnering') and value.

Often, the positive initiatives are supported by legislation and/or contract provisions, such as ISO 9000 certification required as a bidding qualification for government projects in Singapore and Hong Kong or payment timing rules and dispute resolution procedures in the Housing Grants, Construction and Regeneration Act 1996 (England and Wales). However, legislative measures often are initiated to address symptoms rather than causes (or to change culture).

By definition, cultures take a long time to 'brew' and so tend to evolve quite slowly in response to enduring forces. Such forces lie outside the industry as well as within it; indeed, many of the change initiatives have been imposed from outside – whether by legislation and the like as above or by economically or financially driven requirements and actions, as for non-decreasing streams of dividends (Hutton, 2002).

Giddens's (1984) theory of structuration articulates the iterative and integral relationship between social structures, the individuals and groups involved and the mechanisms of changes: people shape social structures (institutions) through their behaviour and, in turn, their behaviour is influenced by the resultant structures – a form of social constructivism; *so too with culture*. An example is the issue of trust: the industry has been depicted as having a 'low trust culture' (e.g. Latham, 1994). Trust has various typologies, one of which is 'dispositional' and 'experiential'. Dispositional trust may be regarded as inherent in an actor and therefore governs the behaviour in a new encounter with a hitherto unknown other. Experiential trust is an aspect of an actor's behaviour towards another based on the actor's experience of previous encounters and/or reputation (probably including rumours too). Thus, if the observation of a 'low trust culture' has validity, it suggests that it results from the natures of the actors and the practices operated in the industry – which, of course, also reflects the, often power-based, impacts of clients and other project participants who are outside the construction industry. Hence, the important questions which emerge concern whether the industry is low in trust; what are the consequences?; what are the objectives of members of the industry and of other stakeholders?; what, therefore, should be changed?; and how should such changes be effected?

Construction projects as joint ventures

A joint venture has become a popular form of formal organization through which construction projects are procured, particularly larger and more complex international projects. Such formal joint ventures see the establishment of a separate business unit by the joint venturing organizations (parents); usually, such joint venture organizations are companies and so the parents set up the joint venture offspring as either an equity joint venture (EJV) or a non-equity joint venture (NEJV) (Glaister, Husan and Buckley, 1998), which

Pangakar and Klein (2001) classify as equity alliances or contractual relationships. Generically, the parents are entering into a business alliance which is 'an ongoing, formal business relationship between two or more independent organizations to achieve common goals' (Sheth and Parvatiyar, 1992).

The issue of 'common goals' should be examined from two primary perspectives: the technical purpose and the business objectives. Although, given reasonably adequate briefing, the technical purpose should be evident, if not totally defined in detail, the business objectives of the alliance partners are likely to be somewhat in conflict, both with each other and with other stakeholders, in the usual zero-sum-game context of distribution of risks and benefits.

The strategic purposes of alliances may be classified as growth, strategic intent, protection against external threats and diversification. However, alliances may be formed for operational purposes: to increase resource efficiency, to increase asset utilization, to enhance core competence or to close the performance gap; a supplementary consideration is that an alliance form may be stipulated as the legally required structure for external organizations to operate in a particular location. Such alliances may be required to provide protective insulation of the domestic organizations and/or to foster technology transfer and organizational learning to the domestic organizations. Alliances with competitors (horizontal) may be made with existing competitors, potential competitors, indirect competitors and new entrants whereas non-competitor alliances (vertical) occur with customers, potential customers, suppliers and potential suppliers.

Especially in construction, formal alliances tend to be entered to enable the parents to participate in projects (transactions) which (otherwise) either are beyond their reach because of size, technical factors and so on or would be external to them. Thus, the necessary formal relationships between the independent organizations which intend to undertake activities together through some pooling of resources lead to the imposition of formal governance structures. That prompts Dietrich (1994) to conclude that formal joint ventures are characterized by oligopolistic partners, which, following Lipsey's (1989: 243) hypothesis of qualified joint profit maximization, stresses the importance of behavioural and consequent performance risks as epitomized under the theory of transaction cost economics.

Sheth and Parvatiyar (1992) determine that uncertainty and trust are the two primary (independent) constructs which affect formal alliance relationships and their institutional arrangements. Bachmann (2001) views trust and power as means by which control is effected within business relationships. Those concerns are commonly manifested in the criteria for selection of partners and the establishment of safeguards against opportunistic behaviour by other alliance members, thereby increasing *ex ante* costs in the venture (Williamson, 1985).

Although a formal joint venture is a particular category of business alliance, informally a joint venture may be seen as a much more generic construct of organizational relationships. An informal joint venture, such as a construction project, is a hybrid of market and hierarchy (firm) elements. The notion of construction projects being, inevitably, joint ventures arises from the realization of projects requiring inputs of several types obtained from various, separate specialists; it is that interdependent essential which demonstrates that the actions of one participant are dependent upon the actions of others and, in turn, impact on the actions of subsequent participants. Such informal joint ventures comprise the more formal inter-organizational relationships (contracts) and the more informal processes through which the individuals working on the project operate and interact to achieve performance (e.g. Tavistock Institute of Human Relations, 1966). That may lead to competing or complementary loyalties and commitments which are likely to have significant consequences for performance achievements and future potential (conflicts, organizational learning, etc.).

Some managerial, administrative or control documents and procedures acknowledge the interdependence, but others operate to foster independence. Fortunately, recognition of the importance of recognizing interdependence and devising processes for actioning such recognition is increasing – such as the observations of Baiden, Price and Dainty (2006) that the procurement methods adopted usually 'have focussed on organisations' individual . . . capability rather than their collective ability to integrate and work together effectively'. Of course, the new version of 'teamwork', namely partnering, should focus on collaborative working but the cautions raised through the findings of, for example, Bresnen and Marshall (2000) and Gruneberg and Hughes (2006) should be noted – it is the real operation of a system which is important to determine the outcomes, not the name given to the processes!

Hence, culture, in all its forms and with all its derivatives, exerts a pervading impact on construction: what is built, for and by whom, how and with what consequences; and so this book.

Overview of contents

Chapter 1 provides an introduction to culture and its important for the construction industry worldwide. It begins by examining the role of construction both as a producer of outputs, mainly as producers' goods and so subject to derived demand, and as a consumer of resources, particularly as a labour-intensive industry. The importance of construction in any economy is considered, including interactions with government as a common, major customer of the industry. The chapter examines construction in relation to the institutional frameworks of societies and the development of social and human capital and proceeds to introduce some primary concepts relating to culture and its investigation. That progresses to review culture in the context of the construction industry and the important conceptualization of construction projects inevitably operating as joint ventures between participants – informally, if not formally too.

Chapter 2 examines construction as a globalizing business, commencing with a brief review of how construction is classified in national and international statistics. The concept of globalization is discussed in the context of construction's role in patterns of development of the built environment and national development. The impact of information technology (IT) is noted with particular attention to its integrating potential for markets, reflecting trends in the financial arena. Regional changes are discussed and issues of ethics, transparency and corruption are examined. Those aspects prompt examination of the identities, roles and impacts of stakeholders, structures and processes regarding how construction projects are realized and the performance (both project management performance and project performance) which is achieved.

Chapter 3 articulates major theoretical bases of culture research and their applications to construction. It commences by addressing the issues of defining culture and proceeds to critique a number of the typologies of dimensions through which investigations of culture have been carried out, at both national and organizational levels. Attention is then directed to examination of cultural differences (and similarities), cultural distance and concerns over changes in cultures. The chapter proceeds to discuss culture in the construction industry, commencing with addressing issues which are widely applicable to the industry's processes and performance. Consideration of alliances addresses their alternative purposes and natures with particular attention to aspects of trust and control. Sustainability concerns include issues of human desires and behaviour in their pursuit, and so prompts debate between 'sustainability' and 'greening' and evaluations of policies and practices. The impact of culture on conflicts and disputes is examined, including why and how they arise and then means for their management and resolution.

Chapter 4 first presents the 3C-Model (Contact–Contract–Conflict) through which the following six detailed case studies of culture on actual construction projects are analysed. The case study projects comprise a complex inner-city project in the Netherlands, a serial housing project in Germany, an infrastructure project in Poland, a factory project in Turkey, a factory design project in the United Arab Emirates and an export market analysis project in China. Each case study is presented and analysed individually to elicit detailed lessons to be learned.

Chapter 5 assimilates the case studies to develop a holistic perspective on the lessons learned. These are categorized as lessons for sustaining contacts, for improving contracts and for prevention of conflicts. The overall lessons are designed to serve as guidance for practitioners in dealing with diverse cultures and cultural issues in project practice, and for academics to help considerations in determining research agendas.

Chapter 6 is the final chapter, which synthesizes the preceding concepts, theories and practical investigations to yield a vision for culture in construction in the future. The chapter commences with a perspective on developments in the industry with focus on information and communications technology (ICT) and on globalization. The critical incidents approach is adopted to address particular aspects relating to common conflicts on projects and to examine developments in human behaviour regarding such incidents. Finally, the nature of business culture is discussed and its role the future of the construction industry; in a world which is increasingly dominated by market economies, that is a vital consideration for the survival and success of construction. How the industry may change is difficult to forecast, but that it will change is certain.

2 Construction
A globalizing business

Introduction

As has been noted in Chapter 1, the definitions of construction, and so what activities are measured and recorded as constituting construction, do vary from country to country and change (slightly) over time. However, the variabilities are a result of history – in particular, the resources of the country and how organized exploitation of those resources has evolved which appear to give rise to 'quirks of classification'. What is noteworthy is that the differences are marginal ones; there is general agreement concerning what are the main activities of the construction industry.

Of course, as activities change over time, both what they are and how they are organized, so the definitions of industries are amended to maintain an accurate reflection of what the mixes of organizations do. The definitions yielding industry classifications are based on functions. How and by whom those functions are carried out, using what resources and with what outputs and other consequences, yield the main data which are collected to produce national industrial statistics.

All industries, by definition, are conglomerates – groupings of different organizations – by type, size and so on. The critical consideration is that, generally, an industry is defined as 'the aggregate of manufacturing or technically productive enterprises in a particular field, often named after its principal product' (Dictionary, 2009). Thus, although the enterprises and their activities are related, usually by common outputs and similar processes, the degrees of similarity (or rather the differences between them) mean that, at the margin, enterprises may be classified into different industries, depending on exact definitions of each industry and how those definitions are applied. The 'Standard Industrial Classifications' give the definitions and, quite logically, are tending to adopt common definitions, typologies and taxonomies – which is essential for valid, and easy, comparisons internationally.

However, the construction industry, classification 'F' (UK and United Nations), is arguably an industry conglomerate with high internal variabilities, both worldwide and within each country. According to the United Nations (2008), Section F is Construction, which comprises subsection 41, Construction of buildings, subsection 42, Civil engineering, and subsection

24 *Construction: a globalizing business*

43, Specialized construction activities. Section L is Real estate activities; subsection M71 is Architectural and engineering activities; technical testing and analysis; M74 is Other professional, scientific and technical activities; N77 is Rental and leasing activities (within N, Administrative and support service activities).

In the UK, the Office for National Statistics (2003) [UK SIC (2003)] also denotes construction as section F but employs a taxonomy which differs from that adopted by the United Nations. Regarding the construction sector, UK SIC (2007) employs the classification of UK SIC (2003).

Beyond the formal classifications employed to produce national statistics, the vital issue is the purpose(s) for which the data are required. That may require reconstruction of the national aggregates, for example to determine the number of persons in an economy whose employment is directly associated with construction (so includes construction firms, architectural practices etc., plant hire and suppliers of materials and goods). Other questions, of much concern to economists, involve determining inter-relationships between industries, to determine dependence linkages of supply and demand (see, for example, Sugden, 1975; Bon and Pietroforte, 1990). Such examinations facilitate understanding of how an industry operates within the context of the national economy and, therefore, the likely consequences of any macro-economic policy instruments directed at that industry. Thus, in the UK and many other countries, the construction industry has, allegedly, been used by government as a medium through which regulation of the national economy has been attempted, as part of a 'package' of fiscal and monetary initiatives; naturally, lead and lag times for such measures to percolate from initiation to consequence can prove highly problematic (for further discussion see, for example, Hillebrandt, 2000).

Beyond purely financial and economic considerations, there are, of course, other important concerns as the livelihoods and welfare of many people are likely to be affected. The processes and consequences (efficiency and effectiveness) are culturally dependent.

Globalization

Globalization constitutes the most recent phase of extension along the continuum of exporting, internationalization and multi-nationalization. Thus, although not a new term, 'globalization' has evolved to indicate a common awareness of a good, service, brand or organization across many (most) countries of the world. Further, it is commonly associated with removal of barriers to movements between nations – whether economic/financial, social or informational. One of the most global activities is access to the internet, which enables awareness of information, activities and so on beyond the user's immediate domain.

An immediate potential consequence of movements towards globalization is convergence between nations and cultures. However, globalization also raises antithetical perspectives to maintain local particularities (language, culture), thereby preserving divergence. That is reflected in certain theories (e.g. Ohmae, 1990), and practices or products (e.g. Coca-Cola), of globalization to yield 'glocalization' (tuning of global products etc. to suit individual, local market demands).

Three categories of infrastructure seem to be vital for globalization, as for other aspects of (economic) development – energy, transport and information – as they facilitate movements and processing of other resources (raw materials, labour and finished goods). Indeed, if we examine any development (instigation, extension, rejuvenation, decline or disposal), the essential presence of appropriate infrastructure is undeniable. Although historically the emphasis has been on real infrastructure – transport, utilities, premises – and this remains of vital importance, today there are additional categories of communications (particularly IT media and facilities) which are taking over roles from transport in many activities (in particular the services sector). The differing roles and natures of infrastructure needed reflect the changing requirement for effective and efficient activities of (potential) occupants/users and, as the pace of change increases, the expectations of future requirements and provision, if only physical space, to accommodate them – in other words, developments require appropriate flexibility.

Physical development around the globe involves a close relationship between production of the built environment and socio-economic (measures of) development. For the construction industry, the pattern is well established; first, infrastructure works must be completed, followed by construction of buildings, and the third category switches emphasis to repairs and maintenance, rehabilitation and refurbishment. Of course, the categories are far from mutually exclusive but denote emphases on the nature of projects and work carried out (demanded and therefore supplied). This has significant implications for the various (sub)sectors of the industry (civil engineering, building) and the importance of sectors of the economy as sources of work (public and private sectors in the world's mixed economies).

Thus, we would anticipate correlation between categorisations of countries in terms of their (Western-oriented) economic development and the mix of construction activities. Employing the common, historic classification of countries as developed, developing and less developed (LDCs) (see, for example, Lipsey, 1989: 742) tends to be emotive, especially owing to its Western development-oriented value judgement; however, it does suggest what categories of construction work are likely to be dominant but gives no indicator of quantities. Today, it is more usual to consider countries as developed (advanced industrialized countries, AICs) or developing and to employ subclassifications of developing countries (newly industrialized countries, NICs; least developed countries, LDCs; see, for example, Ngowi, Pienaar, Talukhaba and Mbachu, 2005).

A more helpful categorization employs categorizing countries by annual income per capita (usually according to data of the World Bank;

World Bank, 2009) as low-income (US$935 or less), lower-middle-income ($936–$3,705), upper-middle-income ($3,706–$11,455) and high-income ($11,456 or more). The calculations for classification relate to gross national income (GNI) per capita according to the methods developed by the World Bank and (in 2009) the classification is based on GNI at 2007. A potentially important problem with such average measurements is the distribution of income – whether fairly even across the population or concentrated within one or more small groups – hence, impacting on spending power and patterns. Income distributions are often assessed from Gini coefficients and from Lorentz curves (see, for example, Samuelson and Nordhaus, 2001: 386–389). There is a tendency for income distributions to become less uneven over time and as development occurs and wealth increases; however, the patterns are not even and are subject to significant impacts from political and social variables as well as economic influences.

The static measures are only preliminary indicators and give a more meaningful picture when assembled as time series to indicate absolute and relative changes. Further metrics are very important too – including life expectancy, literacy, urbanization, population density and international integration. It is notable that Hofstede (1980, 2001) finds correlation between the individualist end of the individualism–collectivism dimension of national culture and measures of economic wealth; he notes impact of longitude also.

Samuelson and Nordhaus (2001) provide a wide-ranging analysis of forms, theories and models of economic development. They determine that three factors emerge as primary influences on economic growth: investment rates, macro-economic management (low inflation and high savings rates) and outward orientation.

For organizations (and their outputs) to become global, the first essential is to have an extensive and secure base in the home market; in some cases, saturation of the home market has meant that the only possible way to grow requires international activity, usually by exporting in the first instance; that often leads, quite rapidly and readily, to localization of sales and marketing leading to the adaptation of outputs, via marginal changes to suit the particulars of local demand (initial 'glocalization'). The next step is likely to be establishing a business unit in the main overseas location(s), but the activities carried out there are likely to be 'production' only, with the main decisions and control remaining in the home country head office. The fourth stage denotes a greater commitment to the overseas locations, including a more enduring presence, by granting much autonomy to the overseas branches (design, financing etc.) and encouraging the employment of local personnel in management, as well as operative, roles ('insiderization'). The final step involves retrenching common activities – branding, financing and so on – back to head office to yield benefits from efficiencies and effectiveness of adopting global perspectives (including issues of suitability and power in the markets) whilst retaining particular details to suit the local customers ('complete globalization' but incorporating 'glocalization'). (For a more extensive discussion, see Ohmae, 1990.)

Especially during the earlier stages of globalization, when risks are of greater potential impact, and the possible negative consequences loom large in corporate decisions, it is common to find organizations entering various forms of alliances to gain a presence in new locations – as withdrawal is much easier. Naturally, alliance forms of business are subject to important issues of commitment, control and distribution (including repatriation) of rewards. It is important to note that it remains fairly common for governments of developing countries to insist, by means of legislation, that all organizations from overseas which wish to undertake projects in that country can do so only if they join with a domestic firm in a joint venture alliance; the objective is to foster the development of domestic organizations through technology and management transfers (i.e. organizational learning). Further, the world's mega projects are too large, and risky, for a single organization to take on alone; thus alliances are usual for mega projects.

Like all new business organizations, new alliances have a high failure rate (about 60 per cent; Anderson Consulting, 1999); Alliance Management International Ltd (1999) attributes 50 per cent of alliance failures to poor strategy and 50 per cent to poor management. Two particular issues for new alliances are compatibility/commonality of objectives for the alliance organization and compatibility of managerial and operational processes; both issues are manifestations of cultural differences between the parent organizations which may lead to power struggles and incompatibilities resulting in the demise of the alliance. Thus, new alliances tend to favour contractual, rather than equity, bonding between the parent organizations, which, to a certain degree, leads to self-fulfilment of the destructive cycle (see, for example, Pangakar and Klein, 2001).

Globalization, in application to project-based industries such as construction, tends to bring organizations from several countries into working contact to undertake specialist roles and tasks on a single project; as in the domestic equivalents, those assemblies of organizations yield temporary multi-organizational coalitions which disperse once their participation in the particular project is completed. What have been widely noted for domestic projects – namely, the communications difficulties and consequent performance problems – are, potentially, far more extensive in the international arena. Thus, although the internet is an increasingly ubiquitous and user-friendly facilitator of global communication, it is only a medium and set of tools; it does not address or resolve the underpinning communications problems of people using different languages, phrasings and so on. However, the more advanced technologies, using multi-media, may help to address some difficulties by providing 'real-time' communications with text, graphics, voice and video which can enable clearer communication and detection and resolution of difficulties at the time.

Information technology (IT) and integration of markets

Many aspects of developments in IT are encouraging globalization through providing remote access to project sites. Such links enable technologists and managers to make decisions from their offices rather than needing to visit the site; information is transmitted via IT media, commonly video links, so that drawings, personnel and parts of the project can be scrutinized 'in virtual reality'.

IT is impacting increasingly on education and training. A number of degree programmes are available through web-based, 'distance learning' initiatives which, over time, are likely to render more traditional distance learning formats obsolete. Books and journals are increasingly available electronically, a few only electronically, and librarians are questioning what the future of their institutions may be in terms of repositories of data and information, access, usage and security. In terms of training people for (physical) skills, technical developments, along with changes in production economics, tastes, fashions and so on, have impacted on what skills are required (and popular). For instance, wet plastering has been largely replaced by dry-lining, hand drafting of drawings has been replaced by computer-assisted design (CAD) packages, engineering structural calculations and design are executed using computer packages – the list of such changes extends hourly.

Given that the use of IT to execute mundane tasks involving repetitive calculation is highly advantageous for elimination of calculation error, and so affords more human resources for non-programmed activities, certain important issues emerge. As computers, once programmed correctly, do not make mistakes – they produce accurate results each time, every time – unless they 'crash', the requirement to check outputs and to adequately understand the processes, including calculations, may be bypassed; the packages become 'black boxes'. Thus, if problems do occur which lead to erroneous output, there is a danger that they may not be detected, or not early enough to avoid major damage and rectification difficulties. As managerial vigilance diminishes rapidly with repeated successes, analogously, vigilance in checking results of calculations, designs and so on diminishes with repeated accurate outputs of results.

As the world's financial markets interact and intermingle, from several perspectives operating as a single global market with a number of diverse physical locations, the locations of suppliers, customers and intermediaries are irrelevant, save for the issues of risk assessment and management and the, quite minor, concerns over exchanges between currencies. More generally, e-commerce is becoming increasingly widespread (eBay, Amazon) and, in construction, e-tendering is used on a few projects.

Thus, for large and prestigious projects, the market for all construction activities is following the trend towards globalization (the 'Three Gorges' project in China being a notable exception). The web provides an important medium for marketing activities and for searching for information to test claims made in marketing literature! Information for marketing, participant selection, bidding, planning, work execution and the like can be made widely available and used. However, the vast majority of organizations in, and associated with, the construction industry are small and tend to lag some way behind the forefront of technological developments; also, their focus is a small, local market. Such firms use IT in very limited ways: accounting packages, estimating, purchasing goods and materials, and some marketing activities. Those firms are necessary followers of trends, the effects of which are 'forced' upon them for survival. They constitute the vast majority of firms numerically, but also are important as employers and as providers of much of the industry's output.

Regional 'integration'

The twentieth century witnessed a number of significant changes concerning several regions of the world. A fundamental underlying change was the reduction of the formal imperialism of, mainly, European countries which had grown over several centuries at the expense, through exploitation, of the subjugated countries. The imperialist powers included Austria, Hungary, Turkey, Italy, France, Belgium, Spain, the Netherlands, Germany and, probably foremost, England (as the UK). The historic exploitation of resources by the imperialist powers concerned natural resources – gold, diamonds and so on – agricultural products and people – most notoriously under the slave trade. The system also worked in the reverse direction through the subjugated countries being captive markets for goods produced by their imperialist masters and sold at controlled prices; the subjugated countries and their trading also provided a continuous source of taxation revenues. Essentially, the end of that imperialist era was marked by the return of Hong Kong to China in 1997 and of Macau in 1998 (although the sovereignty of the Falkland Islands remains an issue for some).

During the twentieth century, the Soviet Union assembled and split, China experienced several changes of regimes and state capitalism ('communism') rose and declined in impact. A notable enduring legacy is the power assembled by the USA, in part the result of two major periods of war (1914–18 and 1939–45), through which the USA became the world's major supplier of many goods and, further, gained huge financial power. The USA also expanded its federal area and, more importantly, its area of economic and political influence to become, effectively, the most powerful nation in the world. Much of that power has been cemented by the immigration of scientists, protection legislation by means of copyrights and patents, and trading legislation to restrict the flows of certain goods, services and information, a good deal of which remains in place.

In part to address the disruptions and destructions caused by wars, the League of Nations was established, which transformed into the United Nations (UN). Membership of the UN has expanded greatly, with many countries joining following their achievement of independence, and its remit of activities has extended into many fields. In finance and economics, the World Bank and International Monetary Fund (IMF) came into existence, along with regional subsidiaries, to assist global cooperation and assistance in financial and economic national and international management.

Regionally, a major development in the later part of the twentieth century occurred in Europe, first with the establishment of the European Economic Community (EEC), the 'common market', initially comprising six countries, through the Treaty of Rome, and the European Free Trade Association (EFTA), initially seven countries. Today, most countries in Europe are members of the European Union (EU), which has evolved from the EEC and covers a greatly extended remit of activities, notably monetary union for many member countries (excluding the UK and some other countries).

A fundamental objective of many evolving international alliances, including the EU and the World Trade Organization (WTO), which China joined in 2001, is the removal of barriers which restrict trade. The EU has laws to that effect and the WTO has membership rules with which member states must comply. The barriers to be removed include quotas and tariffs on movements of goods across national borders as well as restrictions on activities permitted in a country by foreign firms. In the EU, legislation is also in place to allow persons to move freely between member states. Further provisions to encourage free trade relate to encouraging competition – both by requiring international advertising of tenders for projects and not discriminating between tenders from firms in different member countries (EU) but also by having powers to investigate and take actions against firms behaving in such a manner as to restrict free trade (e.g. cartels).

However, contrary policies and instruments do exist in individual countries, some with the agreement of international bodies, which are designed to protect the domestic economy and/or particular industries, often for their development. Thus, for example, Singapore employs an evaluation system for construction projects which allocates weightings to tenders to favour domestic firms; WTO joining arrangements have allowed China time in which some restrictions on foreign firms' activities in China can remain but must be removed progressively.

Several other significant organizations have been formed to facilitate cooperation in international trade and finance, either globally or regionally. Those include the Organisation for Economic Co-operation and Development (OECD), the World Economic Forum and the Association of South-East Asian Nations (ASEAN). Many such organizations are a forum for exchanges of information and debate and, although statements of preferred policies may emerge, the organizations do not have the authority to legislate for their implementation. Similar issues arise for the UN, for example, including requirements for governments to ratify policy and action agreements for them to have legislative effect (such as the Kyoto protocol and the Brundtland Report).

A particular issue of economic policy on competition concerns many construction projects and participant firms – governments and international regulators espouse the virtues of competition whilst, at the same time, extolling the benefits of partnering arrangements (as in PPPs and PFI in the UK). The two, diametrically opposed, messages seem to be that 'competition secures better performance' and that 'enduring relationships secure better performance'; both may be true, along with the possibilities between the extremes, with the most suitable approach depending on the circumstances (i.e. a contingency, or situational, result).

Transparency and corruption

Transparency, or visibility, is becoming an increasingly important requirement of business and other organizational activities around the world. Initially, transparency was emphasized in public sector activities and, in the private sector, through auditing of companies' financial accounts and consumer protection legislation. However, the requirements for transparency are founded in morals and manifested in ethics and social institutions, as well as in the law. These, then, raise issues of accountability, trust and avoidance of corruption.

Transparency concerns developing a system in which the actions are fair and reasonable from the perspectives of all stakeholders and following that system so that all the actions are proper and justified (and can be checked or audited to be so). Normally, the requirements involve producing and following an 'audit trail': the prescribed procedural path to be followed with sufficient evidence being produced so that what has been done can be checked to be correct (accurate and compliant). Naturally, there is considerable subjectivity involved over the morally determined value judgements of what is fair and reasonable (right and proper), including the 'checks and balances' to be incorporated to foster compliance.

The corruption perception index for 2008, produced by Transparency International, placed Denmark, Sweden and New Zealand jointly as the least corrupt (9.3/10) of the countries surveyed and Somalia as the most corrupt (ranked 180 of 180 – 1.0/10); the Netherlands was joint seventh (8.9/10), the UK joint sixteenth (7.7/10) and the USA joint eighteenth (7.3/10). The global corruption barometer noted the perceived increase in corruption, notably bribery, in private sector organizations, but with political parties (followed by countries' civil services) being the most corrupt type of organizations; petty bribery was perceived to be increasing and people did not feel empowered to speak out. Governments remained perceived as ineffective in combating corruption. A press release (Transparency International, 2009) relating to the bribe payers index noted that the construction, real estate, and oil and gas sectors were most prone to corruption; Belgium and Canada were at the top of the index, the Netherlands third, UK fifth, USA ninth, China twenty-first and Russia twenty-second (bottom). (For further details see http://www.transparency.org/.)

Hinman (1997) distinguishes morals, as first-order beliefs and practices about what is good and what is bad which guide behaviour, from ethics, which are second-order, reflective considerations of moral beliefs and practices. Rosenthal and Rosnow (1991: 231) note: 'ethics *refers* to the system of moral values by which the rights and wrongs of behaviour . . . are judged' (italics added). So, ethics are the manifestations of the underlying moral values which guide persons' behaviour through their personal consideration, possibly supplemented and guided by codes of conduct (as produced by professional institutions for their members, or firms' codes of ethics for their employees).

Issues of definition and perspectives on ethics have led to the development of four primary paradigms. Deontology (relating to duty or moral obligation) holds that a universal moral code applies. In scepticism (relativism; subjectivism), ethical rules are arbitrary and relative to culture and to time; that is extended in ethical egoism, whereby ethics are regarded as matters for the conscience of the individual. Thus, egoism concerns pursuit of self-interest and, as such, can be related to common business performance criteria (notably, profit maximization). Teleology (the branch of philosophy relating to 'ends' or final causes) constitutes a utilitarian approach in which ethics are dependent upon the anticipated consequences – leading to a cost–benefit perspective, perhaps invoking the judgemental criterion of 'the greatest good for the greatest number', which itself is likely to necessitate subjectively determined weightings. Objectivism asserts that there are definitions of right and wrong which are accepted generally (either universally or more locally) (Leary, 1991: 261–262).

Given the diversity of ethical paradigms, there remains a lot of scope for variability in deciding what is ethical – a 'tip' (gratuity) in one context or society may constitute a 'bribe' elsewhere. Further significant differences occur across cultures, so the values which underpin the different practices found in different locations promote complexity in understanding and following local behavioural norms (requirements) which are at variance with a person's domestic experience.

Ethics concern how the actions of one person may impact on others, so imposing a 'duty of care' not to harm others. That perspective generates questions of to whom such a duty is owed, together with concerns over whether it should be applied absolutely or relatively (deontologically or teleologically). Those considerations are reflected in dimensions employed to analyse organizational culture: pragmatic/normative (Hofstede, 1994a); universalism/particularism (Trompenaars and Hampden-Turner, 1997).

Law and codes of conduct endeavour to determine boundaries of application (the 'neighbour' principle; the client). National law employs wide boundaries and applies to all persons in the country (the jurisdiction of the law); however, there are many systems of law (including 'common law', 'Roman law', 'basic law' and 'Islamic law') and countries may have more than one system in operation. Codes of conduct usually apply more restrictively and may be specific regarding behaviour towards specified others who are likely to be encountered in the course of activities (notably, the client of a construction professional or consultant). Thus, an objectivist stance is apparent.

Considering that a professional is a person who possesses special knowledge which, itself, concerns generic 'good', and who uses that knowledge for the benefit of the immediate client and wider society, the boundaries of application vanish (see, for example, Koehn, 1994; Pritchard, 1997). Benefit from professional activities for society is the objective but the distributions of such benefits, and of any costs involved, may be left open to judgement because usually a diversity of persons are affected by a professional's work.

The prospect of 'compartmentalization' through the presence of artificial boundaries around behavioural requirements and perspectives – as governed by circumstances (domestic, employment, etc.) – prompts Fellows, Liu and Storey (2004) to discuss the notion of 'personal shielding', in which a person amends his or her (ethical) behaviour to accord with the expressed or perceived ethics of another – usually an employing organization. Such shielding may feature in principal–agent circumstances, including those between a commissioning client and a design consultant. The decision of whose ethics to adopt, and at what level, is important given the common belief of individuals that they have higher personal standards than other people (Ferrell and Weaver, 1978) and than employing organizations. Especially if an organization is able to, effectively, indemnify or compensate employees for consequences of ethical transgressions carried out on behalf of the organization ('in the course of employment'), then it is easier for individuals to adopt the ethics of the organization; that is particularly germane, given the findings of Transparency International, noted above.

Consequences of corrupt actions can be many and varied but usually are detrimental to those not involved in the action. A generic perspective sees social and economic negative consequences as corruption tends to reduce efficiency (including perpetuating inefficiencies) through promoting distortions. Elsewhere, corruption may be used to speed processes or to make something occur – which, although still distorting, could increase efficiency (for the particular matter, if not globally). So, transparency requirements exist to combat corruption but, to be effective, the requirements must be appropriate, enforced and with adequate sanctions for transgressors to motivate compliance. Cultural, moral and ethical variations render design of (universal) transparency systems very difficult – as in the development and use of International Financial Reporting Standards.

Stakeholders, structures and processes

Chandler (1962) determined, from investigation of several large corporations in the USA, that 'structure follows strategy', at least in the long term. However, that perspective should be placed in a more holistic context to

34 Construction: a globalizing business

appreciate generic organizational behaviour. Thus, the fundamental issue is determination of the aim and objectives of the organization, which today are often published through statements of 'vision' and 'mission'; unfortunately, those statements are intended for public consumption and so, frequently, are couched in such generic, vague terms as to be both indisputable (and good) but also incomprehensible in revealing what the organization, and its members, should endeavour to achieve and what to do in operational terms.

Thus, the strategy development processes should translate the vision and mission into statements of objectives for the medium to long term from which targets can be developed to communicate and then gauge short- to medium-term achievements (performance); those can be used as a basis for internal management. The situation is complicated by the dynamism of organizations' environments and the impacts of the organizations' pasts. Thus, strategy must match the variability of the firm's environment, from which it follows that the structure of the organization must be appropriately flexible too. Thus, Figure 2.1 represents organizational constructs which cycle through time to model organizational development (many subcycles are contained within the overall model).

The structures of and processes employed by the world's construction

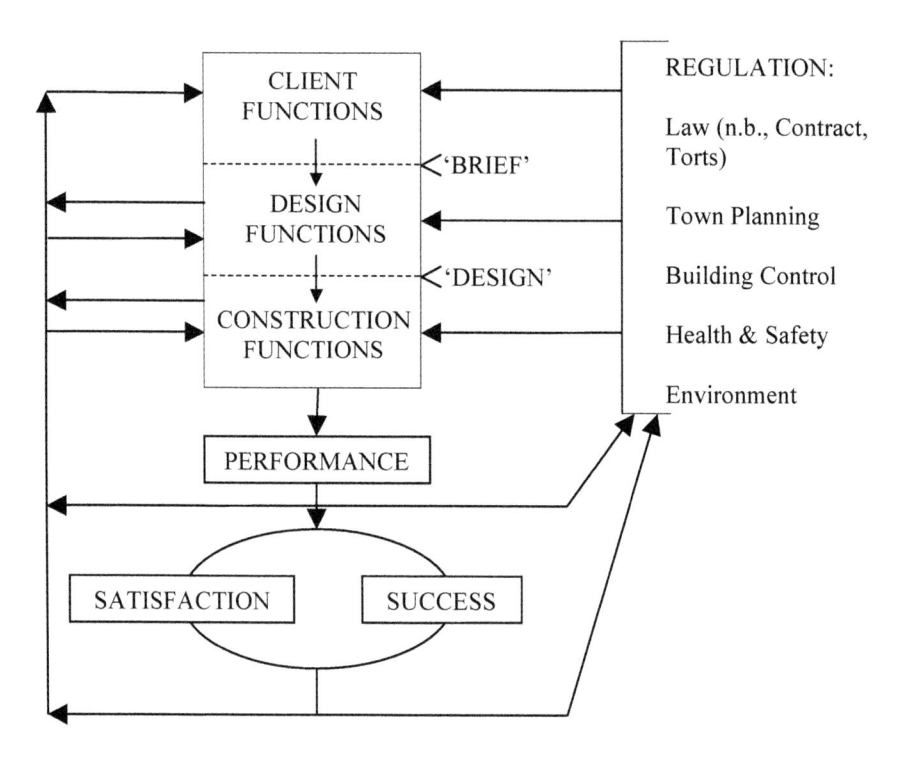

Figure 2.1 Holistic cycle of organizational development. Note: many subcycles exist within the holistic model.

industry to produce the built environment are evolving continuously. Despite the (in)famous observation of Rudyard Kipling (2009),

> I tell this tale, which is strictly true,
> Just by way of convincing you
> How very little, since things were made,
> Things have altered in the building trade . . .

the construction industry in many (developed) countries has changed very significantly in both structure and processes employed in recent years. A major cause of the changes is the recognition of a greater diversity of stakeholders in construction projects and changes in the power (influence) distribution concerning product design and for realization process execution. Figure 2.2 illustrates the project realization process.

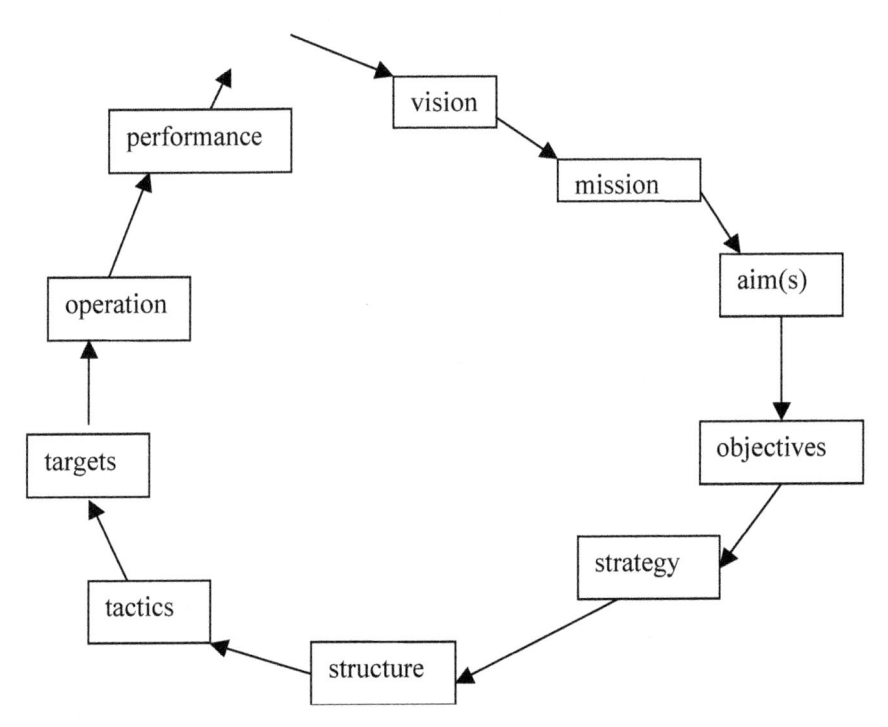

Figure 2.2 The project realization process (developed from Fellows, 2009). Notes: (1) Performance leads to assessment of success and thence to satisfaction of participants and, hence, (perspectives of) overall project success. (2) Performance–satisfaction–success also produces feedforward in the 'cycling' of project data and information to aid realizations of future projects through participants' perception–memory–recall filtering ('experiences'). (3) A similar model applies to projects in use (beneficial occupation) but with 'facilities management' and 'maintenance and adaptation' replacing 'design' and 'construction' as major functionary groups.

Although many members of the industry assert that the environment in which they operate is highly turbulent, which makes prediction and planning much more problematic and, therefore, likely to be subject to large errors, there are facets of the industry's operation which provide some stability. Turbulence can be determined through examination of a number of metrics: notably, orders, output, tender prices and building costs statistics provide ready indicators when considered over time and relative to other industries. Examinations reveal that construction is less turbulent than fashion goods and news media but more turbulent than agriculture, steel manufacturing and other industries which are capital intensive. Also, considering process and costs, construction is notably less turbulent than property development. Internally, building costs (costs to main contractors) are notably more stable than tender prices (prices bid to commissioning clients by main contractors). Finally, the common procedures employed to let construction projects and the durations of projects contribute to certain stability in the short to medium term (predictability of workload) once a project has progressed a certain way in its realization. (An analogous rationale applies in examining turbulence of workload, prices and costs in design activities.)

Thus, perceptions of the turbulence of levels of activity, prices and costs (hence, of profitability) are of great significance in determining suitable structures (physical and financial) of industries and component firms. The more turbulence, the more risk and uncertainty, and so the more flexible the strategy and structure required for survival and success. That indicates that the boundary must be highly permeable to facilitate environmental scanning and rapid, appropriate response to the changes predicted. Construction firms must be structured and operate as open systems, with permeable boundaries and flexible internal mechanisms, to survive.

Stakeholders in a major construction project are likely to extend beyond the primary groups which are concerned with almost all construction projects to include interest and political (pressure) groups – notably those concerned with the environment, particular social aspects (e.g. community preservation) and the like. Identification of stakeholders on any project is important, to determine the immediate concerns and considerations which are likely to impact on the project during its realization, during its subsequent use (including maintenance and adaptations) and during its disposal.

Stakeholders shape the landscape of projects (the macro level of types, mixes, sizes, locations) as well as impacting on individual projects. Adopting a functional perspective, as in Figure 2.1, the client stakeholders' primary functions are commissioning the project, financing, owning and – particularly – using the project (as an output product). Design stakeholders' functions comprise architecture (and its associated aspects of landscape and interior design), structural (civil) engineering, services (environmental) engineering and quantity surveying (value management and cost engineering). Construction stakeholders mainly include main contractors, subcontractors, and suppliers of materials and components; this group is likely to include many diverse firms on any single project.

On any construction project, various participants are involved in instigating, designing, constructing, occupying and using, and finally disposing of the project. All those activities occur in a context of diverse interest groups and of official regulators. Each participant has particular interests and objectives for its own survival, development and business success, and tries to pursue through participation in the project. A simplified, functionally oriented approach to the primary activities of designers and constructors of construction projects indicates their primary interests to be:

- architect – arrangement of spaces based on aesthetics;
- structural engineer – structural integrity;
- services engineer – human comfort;
- quantity surveyor – cost minimization;
- main contractor – profitable construction of the design.

Those functions operate within the usual, overall objective of 'satisfying the client'. Hence, further important questions arise:

- Who is the client?
- How to determine what is required to satisfy the client?
- How are the different, primary interests of the major participants reconciled?
- What are the likely effects on project performance (the product; the project in use) and on project management performance (the project realization process)?

As projects increase in size, so complexities increase geometrically through technical inclusions, scale, diversity of specialists and, consequently, managerial requirements – coordination, especially.

The participants operate in business contexts which are increasingly market oriented. Usually, work (revenue) is obtained through 'winning' projects by competitive bidding; then, for the organizations to survive, the revenue must produce enough profit to provide the owners of the businesses with sufficient returns on their investments. Thus, the imperatives of the business environment place project participants in potential conflict over cash flows and revenue distributions, which tend to be reinforced by differences in the specialist, technical objectives.

As objectives drive behaviour and, thus, performance, it is essential to ask 'what determines objectives?' The answer lies in people's values and beliefs, including their moral systems, the fundamentals of culture.

The activities of the client, design and construction stakeholders are regulated by, and so subject to the requirements of, the increasing number and authority of regulators – usually part of government, or governmental agencies, which are charged under statutes with scrutinizing and assessing designs and realization activities to ensure that they comply with legislation (planning policy and regulations, building control, health and safety, environmental protection, etc.). Such legislation is 'protective' of many and diverse aspects of 'the public good'. Initiated as measures to protect the health of the community, the scope has extended to include safety, resource preservation and environmental protection as well as securing a desirable (hopefully, effective and efficient) built environment – at both the macro 'cityscape' and individual building levels. Much of the legislation is moving away from prescriptive content of what can and cannot be built, via performance desirables and 'deemed to comply' specifications, to the approach of using performance principles against which designers of buildings must demonstrate compliance.

Certain regulatory functions are performed by agents who operate in the private sector. Various functions have been privatized in some countries whereas others, such as certifications under ISO systems, have always resided in the private sector. Thus, for clients, designers and constructors, it is essential to comprehend the regulatory system and its requirements which apply to each project to ensure full compliance with the applicable, local law – do not assume that what applies in the home country (or even locality) applies elsewhere and in the same manner.

In analysing the requirements and desires of stakeholders, it is helpful to adopt a four-step approach (as widely used in risk management – see, for example, Fellows, 1996): identification, quantification, allocation and response. Identification comprises two stages to determine who the stakeholders are and what are their criteria (positive and negative) pertaining to the project; this may require establishing a 'threshold of influence' to ensure that only stakeholders and criteria which are likely to be significant are included. That can be problematic as value judgements are required and so it is important to be transparently objective. Quantification is always difficult as each stakeholder is likely to emphasize both the absolute and the relative importance of his or her group and its criteria. Equity, to deal with all stakeholders and criteria fairly, is extremely difficult in these value judgements. Again, endeavours to be objective are assisted by visibility. The quantifications are, preferably, in the form of ratings of the importance of each stakeholder, and of each criterion; rankings alone are problematic as they express only relative positions (ordinal scales) with no indication of distances between ranks – ranks can, of course, be produced from ratings but not vice versa! Allocation concerns deciding which participant(s) in the project is most suitable to deal with the criteria; many will be design aspects and require inputs from various specialists. The final stage, response, involves addressing each criterion, often iteratively, to secure an outcome which is appropriate to the criterion and compatible with the other aspects of the total project. The iterations may occur not only by the project participants producing an initial response but through cycling of proposed responses and reactions by stakeholders until a workable project emerges. That final stage may involve much 'political decision making' (as discussed by, for example, Hall, 1981).

A further complication arises as some stakeholders are likely to be represented by agents, for certain purposes, if not entirely. Thus, the agents are likely to act as 'filters/interpreters' of the principals' (stakeholders') requirements – such as the common role of the architect or engineer as the agent of the employer (client) for specified purposes under standard forms of construction contract. Clearly, an extensive communications network of individual communications chains is likely to be involved, thereby geometrically increasing the potential for misunderstandings and errors (including agents incorporating their own criteria).

The generic trend is for the number and diversity of stakeholders to increase over time as projects, and their related activities, increase in complexity and so disciplines spawn specialist branches. However, concerns over the negative consequences of fragmentation have fostered some integrations, such as multi-disciplinary design practices, design and construct organizations, and developers which also design and construct (usual in 'speculative housing').

Occasionally legislation, but usually perceptions of opportunities and threats (business potential gains, and risks and uncertainties), prompt structural changes in the industry. Financially based power, as impacted by economic cycles, has shifted the distribution of power on construction projects towards clients, who, thereby, can be more active in expressing demand (their performance requirements) rather than accepting what the industry decides to produce. Such changes in performance, process and output products are important forces in shaping the structure of the industry. Hence, the rise of concession arrangements, in their great variety of guises – including PPP and PFI. In the construction contracting arena, risk distributions, returns on investments, employment legislation, taxation regulations, workload variability and so on have promoted a major and widespread structural change: traditional 'main contractors' do not exist in many countries; they are now management contractors who organize work package execution by a diversity of numerous subcontractors. In some instances (e.g. the UK), that change is supplemented with changes in subcontracting mechanisms away from designer/client selection (as under 'nomination') to contractor selection (domestic). That change adds to the trend to give (more) responsibility for design to the contractor (constructors).

At the international and global level, structural changes in the industry not only reflect the national trends of changes in structures and processes but also, notably, demonstrate trends in 'big business', especially reflecting the impact of the growth criterion. Through various forms of amalgamations and take-overs, many construction firms are vanishing as independent business units; the number of large firms in the industry is diminishing whilst, at the same time, the size of the remaining large firms is increasing. Important elements of that phenomenon are noted by Furnham (1997): 'Usually the corporate culture of the most powerful or economically successful company dominates.'

The changes indicate that the world's economy, reflecting national economies, is increasingly dominated by a declining number of increasingly larger, conglomerate corporations; construction is no exception. It is noteworthy that the world's highly dominant economic system, market capitalism, operates in a large number of variants – Dutch capitalism differs from Swedish, British, American, Hong Kong, Chinese and other versions – so the national and organizational cultures (and climates) are important in order to understand the drivers and motivators of objectives, strategies, structures and operations, as well as changes therein.

Within China, a major economic power and likely to become dominant, amalgamations amongst construction conglomerates have resulted in the China Railway Group becoming the largest construction organization in China, taking over from China State Construction. In 2009, four of the largest ten construction contractors in the world were Chinese (International Construction, 2009). Notably, despite major changes in the China economic system, most major organizations remain as state-owned enterprises of various forms. (For a more detailed discussion, see, notably, Hutton, 2006.) Even before the 'open door' policy was implemented, Chinese contractors were operating overseas, particularly in Africa. Since the opening of China, that country has become a major consumer of international construction, including design, and materials and components, as well as an important provider of construction services in the world market (but with particular emphasis on certain locations: Africa, the Middle East). The modes of operation of Chinese firms overseas is changing also, from being a tightly closed system (all personnel, materials and methods brought from China and contained within a difficult-to-penetrate physical boundary around the site) to a more open system. (The nature of the system adopted depends on the project and its location.)

Apart from the burgeoning of PPP projects (commonly due to financial concerns of the public sector clients), formal joint venturing is becoming more extensive. The sizes and perceived risks of the largest projects around the world often are too great for any single constructor to take on, so consortia are formed. Elsewhere, as an enduring policy amongst developing countries, there are legislative requirements for overseas constructors to form a joint venture with at least one domestic constructor to be permitted to work in the country. Two important aims of such requirements are to secure technology and management skill transfers to the domestic firm(s). The other main rationale for joint venturing is to secure a place in a new market (locationally, horizontally or vertically) at a lower risk than lone entry (such as by setting up a subsidiary) to that market; the risk is reduced by utilizing the market knowledge of the local, established partner. An important component of such risk is culture, which influences local business processes and practices and much of which is likely to exist in the form of tacit knowledge and so be invisible to people outside that culture – as discovered by Japanese car producers, in establishing plants in the USA, ignoring local culture and its business manifestations can be very detrimental to performance (Womack, Jones and Roos, 1990).

Further process changes reflect the procurement dichotomy between performance improvement perceptions of competition and cooperation. During the early years of the twenty-first century, 'framework agreements' have been developed to attempt to combine enduring relationships between commissioning clients and main contractors – suitable where the client has a prolonged or continuous programme of work (which may vary considerably in the short term). 'Framework agreements are agreements with one or more suppliers which set out the terms and conditions for subsequent procurements' (Office of Government Commerce, 2006). Framework agreements may be contracts or non-contractual, with one or more suppliers, and are used in both public and private sectors (thus, in some cases, framework agreements resemble measured term procurement). Often, the intention is for similar framework agreements to be put in place down the supply chain (i.e. with subcontractors and suppliers) to generate a network. 'Call-offs' (individual procurements under the framework) are usually effected through tendering for that provision by the firm(s) in the framework. Thus, framework agreements are designed to combine elements of 'partnering' in the framework with competition in the call-offs. What performance finally results remains undetermined objectively.

3 Culture's influences in construction
Theory and applications

Introduction

Late in 1993 the interim report of Latham's review of the UK construction industry accentuated major problems of lack of trust and lack of money (Latham, 1993). The following year, the final report was published (Latham, 1994) but, given its more extensive content, especially concerning payment terms and practices and the call for 30 per cent improvement in productivity, much of the message regarding trust was obscured.

The Egan Report (Construction Task Force, 1998), *Rethinking Construction*, suggested the creation of a 'movement for innovation' and, in so doing, focused on five 'drivers of change': committed leadership, a focus on the customer, integrated processes and teams, a quality-driven agenda, and commitment to people.

Although the reports are UK-centric, the investigations incorporated findings of studies in USA and Europe as well as a continuous 'eye to' Japan; considerations include practices and performance in a variety of other industries, with the practices and achievements of Toyota being clearly influential. Resultant recommendations and performance target suggestions tend to take the form of command and control, as in 'drivers of change', and so originate outside the construction industry. While such 'demand–pull' forces are powerful, they must be transmuted into incentives to enduringly motivate the industry internally for take-up to be widespread and sustained – by effecting the desired cultural changes expressed and implied by the reports.

Certainly, there is much room for improvement in the components, practices (processes), products and hence performance of the construction industry in all countries. However, the all-important questions concern what to change, how, and what the consequences will be; the 'why?' as an overall question is very important and, usually, relates very closely to 'who' considerations. Simplistically, the problems' causes are concentrated in fragmentation and adversarialism – as manifested in the ubiquitous obsession with competitive tendering for work allocation; the 'lowest bid wins' approach not only is naïve, especially for value achievement, but immediately establishes a situation which is highly conducive to opportunistic behaviour and its close associate, adversarialism.

One common cry is that the extensively detailed contracts, often the industry's standard forms, are adversarial documents – indeed, the Egan Report, through its promotion of partnering, suggests that pursuit of partnering could lead to the disappearance of contracts! That perspective raises two important issues. One is that, even if no formal contract has been executed, English law, and legal systems based upon it, will presume that a contract was intended (from conduct etc.) in business circumstances unless there are express indications to the contrary. The second is that the contracts, in themselves as legal documents, are not adversarial but the people who use the contracts and the processes they employ may be adversarial. (Confusion arises because many of the standard form contracts used in the construction industry incorporate a large number of procedural terms – if 'X' happens, then party 'A' does α, β, γ.)

It is widely acknowledged that two systems exist concurrently on construction projects: the formal system as laid out in the contract procedures, organization chart and so on; and the informal system, which is a function of the people working on, and in close connection with, the project and the relationships between them. Indeed, a common perspective is that projects must operate through the informal system, and to the exclusion of the formal system, in order to be effective and efficient in meeting the performance targets; close adherence to the procedures (times etc.) of the formal system would lead to delays and cost overruns inevitably. Thus, the formal system is largely ignored unless/until things go so wrong on the project that the formal, legalistic system is invoked (i.e. conflict escalates into dispute) to obtain redress and achieve a solution. (See, for example, Tavistock Institute of Human Relations, 1966.)

So, essentially the problem causes are people, their processes/practices and relationships: functions and manifestations of culture.

Therein lies a major part of the solution: understanding culture and how change may occur. Figure 3.1 illustrates a simple, generic model of culture.

Culture: definitions and nature

An initial description of culture is 'how we do things around here' (Schneider, 2000). However, much more is involved: culture is not just *how* things are done but a far more extensive construct; culture includes *what* is done, *why* things are done, *when, by whom* . . .

Kroeber and Kluckhohn (1952) found 164 definitions of culture; the count is likely to be considerably higher today. They define culture as:

> patterns, explicit and implicit, of and for human behaviour acquired and transmitted by symbols, constituting the distinctive achievements of human groups, including their embodiment in artefacts; the essential core of culture consists of traditional (i.e. historically derived and selected) ideas and, especially, their attached values; culture systems

44 *Culture's influences in construction*

may, on the one hand, be considered as products of action, on the other as conditioning elements of future action.

Culture is a construct which concerns groups of people, rather than relating to particular individuals (as personality does). Further, culture is iteratively dynamic: culture shapes behaviour and, in turn, behaviour shapes culture; therefore, development of culture spirals through time.

Hatch (1993) employs a model of cultural dynamics which incorporates the cyclical processes of manifestation, realization, symbolization and interpretation in the development of cultures. The dynamism comes from the continual formation and reformation of a culture as the context for constructing images, forming identities, determining meaning, setting goals and taking action. In construction, the impact of culture on what performance is achieved and measured against predetermined, culturally bound targets is important in assessments of project (and organizational) success and stakeholder satisfaction.

Hofstede (1994b) defines culture as 'the collective programming of the mind which distinguishes one category of people from another.' That definition indicates that culture is learned, rather than being innate in people or

Behaviour, Heroes, Symbols, Artifacts, Language, etc.

Values

Beliefs

Practices

(Fundamental)

(Hierarchy)

(Manifestations)

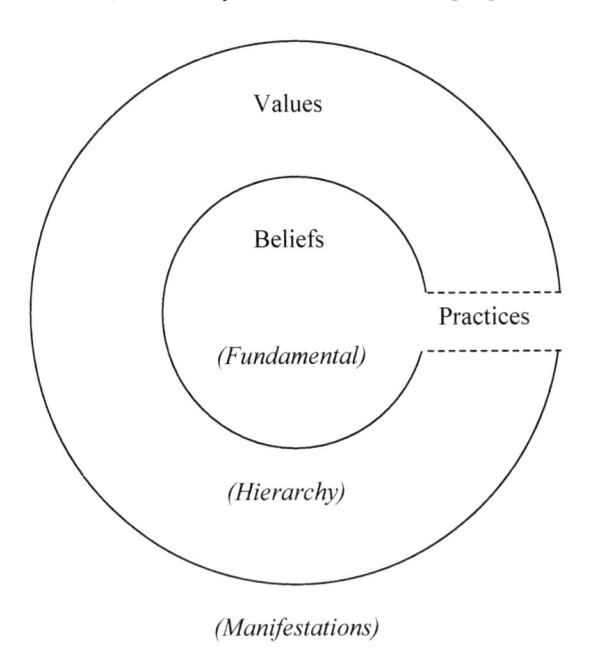

Figure 3.1 Layers of culture (derived from Hofstede, 1980, 2001). Note: Schein (2004) considers levels of culture to be artefacts, espoused beliefs and values, and underlying assumptions (as the deepest level).

inherited genetically. However, culture is inherited behaviourally through perceiving, replicating and responding to the behaviour of others.

As culture is a collective construct, categorization of people may be by ethnic origin, political nation, organization and so on. Normally, 'within-category' variability should be significantly less than 'between-category' variability. That is very important in examining similarities and differences between cultures – where the boundaries are drawn (for 'national' cultures, the existing geographical border of the nation is the usual boundary), what dimensions are considered and how they are measured. The other common boundary used in investigating culture is organizational: a company, government department, professional institution, university or some other appropriate grouping of persons.

Schein (1990) believes that the essence of culture is a pattern of basic assumptions which constitute communal values which are taken for granted by the persons constituting the cultural group (so, often, are not expressed in documents such as mission statements). Thus, cultures arise, first, through the formation of norms of behaviour relating to critical incidents (often as lessons learned from significant mistakes) which are commonly communicated through stories passed on between members of the community (in the form of national or organizational histories) and, second, through identification with leaders and what they scrutinize, measure and control.

Behaviour is a primary manifestation of culture. Behaviour depends upon attitudes, values, beliefs and assumptions – whether the behaviour is determined by conscious thought or evaluation or is 'instinctive'. Common survival mechanisms govern much instinctive behaviour and, therefore, are relatively consistent amongst people; for behaviour determined by thought (cognitive components), cultural influences are stronger. Such perspectives yield models of culture which employ concentric analyses, with physiological instincts and beliefs at the core (survival imperatives, religion, morality etc.), values as the intermediate layer (the hierarchical ordering of beliefs, perhaps with possible trade-offs) and cultural manifestations at the outer layer (as in behaviour, language, symbols, heroes, practices, artefacts), as in Figure 3.1. Culture may be subject to groupwise analysis and categorization as well. That approach includes the categories of national culture, organization culture, organization climate, and behaviour of persons (see Figure 1.1).

The boundaries of the categories are quite fuzzy and the categories 'bleed' into each other. Individuals, at any one time, may be of a certain nationality (e.g. Chinese), work in a particular organization (e.g. Dragages), belong to a 'social interest group' (e.g. Greenpeace), and exhibit certain behaviour (e.g. organizational citizenship behaviour). Additionally, situational variables (circumstances) impact on the behaviour, which, then, can be regarded as somewhat contingent (see, for example, Fiedler, 1967). Such complexity has prompted a variety of analyses for each of the categories. One consequence is confusion of the categories when 'managers' wish to effect change – in particular, what has been changed and the strength and permanence of any

46 *Culture's influences in construction*

change; culture cannot be changed by use of a '40-hour workshop' (although behaviour modifications may result).

Eldridge and Crombie (1974) note three concerns for culture: depth (values and commitment), breadth (coordination of the persons) and progression (coordination, development/change, over time). Often, cultures are caricatured by general perspectives which are used to serve as (simplistic) guides to how people in those cultures behave and how visitors are expected to behave (see, for example, Trompenaars and Hampden-Turner, 1997).

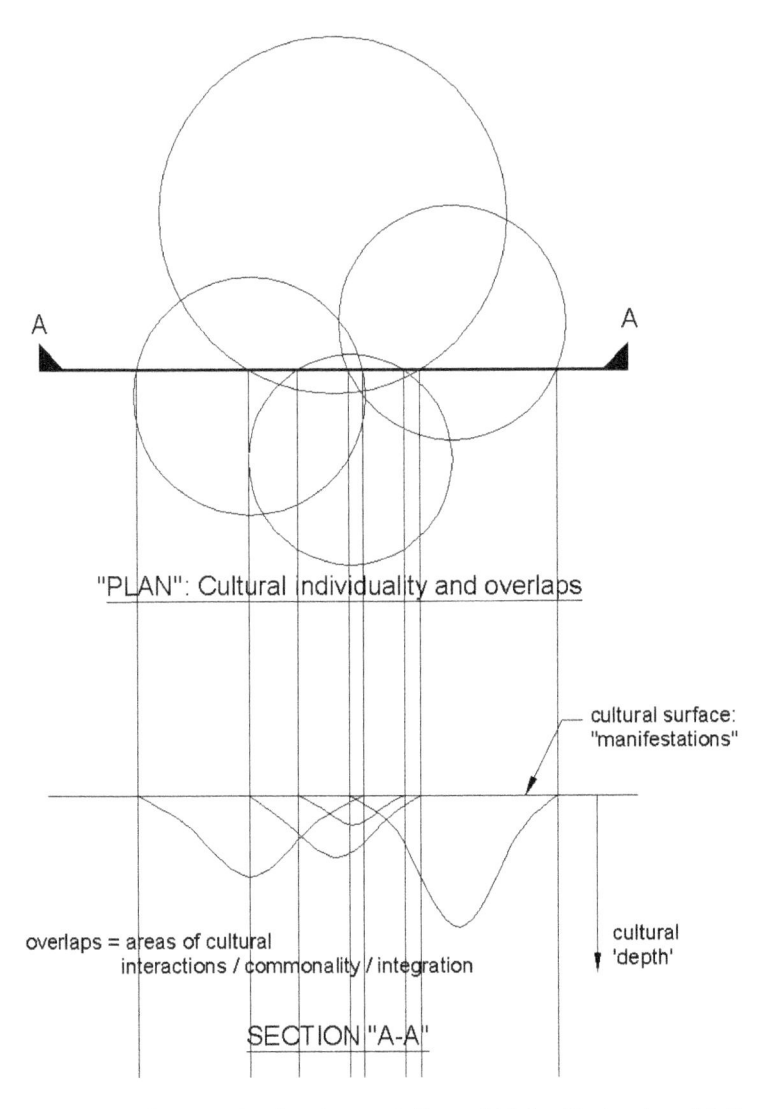

Figure 3.2 Inter-relations of (sub)cultures (Fellows, 2006b: 52, developed following Trompenaars and Hampden-Turner, 1997).

Figure 3.2 portrays how cultures or subcultures may overlap and interrelate to each other. Commonly, people must move between cultures and so, to varying extents, adapt to the cultural context which they perceive to prevail. Generally, some attention has been directed at more enduring, obvious moves between cultures, such as secondment of staff to projects overseas. The concerns relate to selection of appropriate staff (and families) and suitable training for them. However, once transferred to the new culture, there is a normal pattern of development – acculturation – which occurs over a considerable time (usually a few years), as depicted in Figure 3.3.

Recently, there has been much interest in comparisons of the general characteristics attributed to Eastern societies and people in comparison with those in the West. Eastern people are often viewed as 'fatalistic' (or 'flexible' or 'adaptive') whereas Westerners are characterized as would-be controllers. Traditionally, Eastern people tend to regard themselves as subservient to natural forces and desire to be 'in harmony' with nature (as do people from other parts of the world, such as Africa and Australia); change and, hence, uncertainty are accepted as inevitable, truth is determined by spiritual (religious and philosophical) principles and considerations of time are long-term. Western people, in contrast, tend to regard themselves as somewhat in control of nature and therefore able to harness many of its forces to improve human society; so, for Westerners, time, change and uncertainty can be managed; truth has a scientific base and so is determined by measurements and facts.

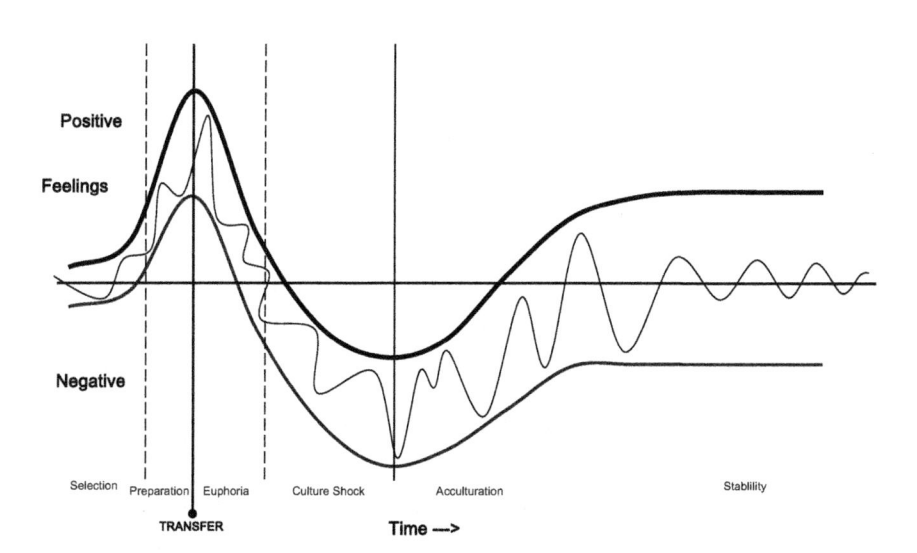

Figure 3.3 Usual pattern of acculturation–adaptation to a new cultural context (Fellows, 2006b: 59, developed following Hofstede, 1994a).

Dimensions of cultures

Introduction

Culture may be studied through collecting and analysing data relating to its manifestations: behaviour, language, symbols and so on. Such data and analyses may be used to help understand the deeper levels of cultural, latent constructs: the values and beliefs (or other constructs depending on the model of culture adopted) of the members of the society under study. Measurements are often comparative and so yield relative, rather than absolute, results (e.g. Hofstede, 1980).

Languages may be analysed in terms of how they express meaning: high content or high context. Hall and Hall (1990) employ the concept of high context versus low context as a primary dimension along which entire cultures may be analysed, as demonstrated in Table 3.1. In a high-context

Table 3.1 High-context/low-context (high-content) cultures

Factor	High-context cultures	Low-context cultures
Overtness/precision of messages	Many covert and implicit contents of messages, with much use of metaphors and 'reading between the lines'	Overt and explicit messages that are simple, clear and precise
Locus of control and attribution for failure	Internal locus of control and personal acceptance of failure	External locus of control and blame of others for failure
Use of non-verbal communications	Many non-verbal communications	Focus on verbal communications
Expression of reactions	Reserved, inward reactions; few external signs	Visible, external, outward reactions
Cohesion and separation of groups	Strong distinction between in-groups and out-groups, belong to few in-groups; strong sense of family	Flexible and open grouping patterns, changing as needed; may belong to many groups
People bonds	Strong people bonds with affiliation to family and community/in-groups	Weak bonds between people with little sense of loyalty
Level of commitment to relationships	High commitment to long-term relationships; relationships more important than tasks	Low commitment to relationships; tasks more important than relationships
Flexibility of time	Time is open and flexible; process is more important than product	Time is highly organized; product is more important than process

Derived from Hall and Hall (1990).

culture, there are many contextual elements that help people to understand what messages, behaviour and other manifestations of the culture mean. In such cultures, a lot is taken for granted, so meaning must be derived from the message itself and its interpretation in the prevailing circumstances or situation; a great deal of intuition is necessary together with a thorough understanding of both the language and the society. Such environments can be very difficult and problematic for people from other cultures. In a low-context culture, very little is taken for granted. Thus, much more content is needed but the resultant message is precise and explicit in its meaning, so there is a low chance of misunderstanding; thus, people can be quite confident to interpret messages at 'face value', although such direct and obvious expression can be offensive to people from high-context cultures.

Hall and Hall's second dimension concerns how people perceive time along a dimension of monochronic versus polychronic (see Table 3.2). Monochronic time means doing one thing at a time, often in a predetermined sequence. More complex situations (such as a construction project) require careful planning and scheduling to be carried out, and this is usually achieved by practising 'time management'. Monochronic people tend to be low context (high content). In polychronic perceptions of time, human interaction is valued over time itself and material things. That generates a low concern for 'getting things done' – they get done, 'in their own time'. Polychronic people tend to be high context.

Trompenaars and Hampden-Turner (1997) examine cultures' approaches to time along a dimension of sequential versus synchronic (corresponding to monochronic versus polychronic). Further, they examine how people of different cultures consider time as periods – past, present and future – regarding the relative importance of each period and how the periods relate to each other (overlaps or distances of separation). Those relative importances of

Table 3.2 Perspectives regarding time

Factor	Monochronic action	Polychronic action
Actions	Do one task at a time; sequential activities	Do several tasks simultaneously; parallel activities
Focus	Concentrate on the task in hand	Easily distracted
Attention to time	Emphasize when things must be achieved	Emphasize what will be achieved
Priority	Tasks	Relationships
Respect for property	Seldom borrow or lend things	Borrow and lend things frequently and readily
Timeliness	Emphasize promptness	Base promptness/timing on effects on relationships

Derived from Hall and Hall (1990).

periods relate closely to Hofstede's fifth dimension of national cultures: long-termism versus short-termism.

The third dimension of Hall and Hall relates to territory, or space: whether persons are highly possessive over space and objects. High territoriality includes clear and firm demarcation of (often physical) boundaries with high needs for security to protect ownership (rights). People who are highly territorial tend to be low context (high content) and are likely to desire large amounts of physical space. For those who have low territoriality, space, possessions and boundaries are low in importance and so, for them, movements are easier. Low-territoriality people tend to be high context.

Hofstede (1980) determines four dimensions from studying national cultures:

- power distance – 'the extent to which the less powerful members of institutions and organizations within a country expect and accept that power is distributed unequally' (Hofstede, 1994b: 28);
- individualism/collectivism – 'Individualism pertains to societies in which the ties between individuals are loose: everyone is expected to look after himself or herself and his or her immediate family. Collectivism as its opposite pertains to societies in which people from birth onwards are integrated into strong, cohesive in-groups, which throughout people's lifetime continue to protect them in exchange for unquestioning loyalty' (ibid.: 51);
- masculinity/femininity – 'masculinity pertains to societies in which gender roles are clearly distinct (i.e. men are supposed to be assertive, tough, and focused on material success whereas women are supposed to be more modest, tender, and concerned with the quality of life); femininity pertains to those societies in which social gender roles overlap (i.e. both men and women are supposed to be modest, tender, and concerned with the quality of life)' (ibid.: 82–83);
- uncertainty avoidance – 'the extent to which the members of a culture feel threatened by uncertain or unknown situations' (ibid.: 113).

A fifth dimension of long-termism – 'the fostering of virtues orientated towards future rewards, in particular perseverance and thrift' (ibid.: 261) – versus short-termism – 'the fostering of virtues related to the past and present, in particular respect for tradition, preservation of "face", and fulfilling social obligations' (ibid.: 262–263) – has been added (Hofstede, 1994a) following studies in Asia using a Chinese values survey (CVS) which detected important impacts of 'Confucian dynamism' (Chinese Culture Connection, 1987).

In 2010, based on Minkov's World Values Survey data analysis for 93 countries, a sixth dimension was added, indulgence versus restraint. Indulgence concerns a society which 'allows relatively free gratification of basic and natural human drives related to enjoying life and having fun.

Restraint stands for a society that suppresses gratification of needs and regulates it by means of strict social norms' (Hofstede, 2011).

Hofstede's research and the dimensions of culture (both national and organizational cultures) which have been determined from it remain the best-known and most widely used models and dimensions. Despite some notable criticisms (e.g. McSweeney, 2002), the work remains widely employed in both academic research and practical application – constituting evidence for its robustness.

Chen, Meindl and Hunt (1997) examine the division of the cultural construct of collectivism into vertical and horizontal components. They juxtapose those components to Hofstede's (1980) dimension of individualism, as 'individualism (low concern for collectivity and low concern for in-group others) at one end of the spectrum with vertical collectivism (high concern for the collectivity) and horizontal collectivism (high concern for in-group others) at the other end'.

Triandis and Gelfand (1998: 119) extend that analysis, arguing that both individualism and collectivism have horizontal (emphasizing equality) and vertical (emphasizing hierarchy) components. They assert that 'the most important attributes that distinguish among different kinds of individualism and collectivism are the relative emphases on horizontal and vertical social relationships'. Horizontal individualists (HI) want to be unique and distinct from groups, for example 'I want to do my own thing' (e.g. Norway), whereas vertical individualists (VI) want to be distinguished, acquire status and compete with others, for example 'I want to be the best' (e.g. the USA). Horizontal collectivists (HC) emphasize common goals, interdependence and sociability (e.g. Israeli kibbutz), whereas vertical collectivists (VC) emphasize the integrity of the in-group and are willing to sacrifice their personal goals for the sake of in-group goals (e.g. China).

Chen and colleagues (1997) note that, 'Because the vertical scale items refer to work situations and the horizontal scale items primarily refer to non-work situations, one may speculate that the Chinese are becoming "organizational individualists" even though they are still cultural collectivists in other domains'. That change, which may be generalized to include the 'Asian Tiger economies', is probably due to the rapidly rising levels of industrialization and wealth in those countries (e.g. Hofstede, 1983, 1994a: 75; Triandis, 1990). Hofstede (1983) notes the correlation between wealth and individualism in various countries and continues that 'Collectivist countries always show large Power Distances but Individualist countries do not always show small Power Distance'. Gomez, Kirkman and Shapiro (2000) explain that people in collectivist cultures favour in-group members but discriminate against out-group members.

Trompenaars and Hampden-Turner (1997) advance five value-oriented dimensions of culture which, they suggest, 'greatly influence our ways of doing business and managing as well as our responses in the face of oral dilemmas'. Those dimensions, which although pertaining to national culture seem particularly relevant to organizational analyses, are universalism–particularism (rules–relationships), collectivism–individualism (group–individual), neutral–emotional (feelings expressed), diffuse–specific (degree of involvement) and achievement–ascription (method of giving status).

Organizational culture

Combinations of cultural manifestations, especially language and behaviour (to emphasize the 'deal', the organizations, or the interpersonal relationships), have major impacts on whether a deal is struck, with whom, using what formal and informal frameworks, how it is executed and with what consequences (see, for example, Trompenaars and Hampden-Turner, 1997). To carry out business successfully, particularly international business, being appreciative of and sensitive to behavioural differences – and their reasons and consequences – is vital for success.

Many of the difficulties which remain common on domestic as well as on international construction projects seem attributable to two primary causes – conflicts between the business objectives of participants (often due to the zero-sum-game nature of the (standard) processes used, notably competitive tendering), and lack of sensitivity to and accommodation of differences between participants; hence the need for awareness and understanding of organizational cultures.

Hofstede (1994a) defines organizational culture as 'the collective programming of the mind which distinguishes the members of one organization from another'. He proposes six dimensions for analysis of organizational cultures:

- process–results orientation (technical and bureaucratic routines [can be diverse] versus outcomes [tend to be homogeneous]);
- job–employee orientation (derives from societal culture as well as influences of founders and managers);
- professional–parochial (educated personnel identify with their profession[s]; people [also] identify with their employing organization);
- open–closed system (ease of admitting new people and of innovating; styles of internal and external communications);
- tight–loose control (degrees of formality, punctuality, etc.; may depend on technology and rate of change);
- pragmatic–normative (how to relate to the environment, notably customers; pragmatism encourages flexibility).

Usually, organizational cultures are initiated by the founders of the organization and are amended by others who have had a major impact on the organization's development (e.g. Henry Ford, John Harvey-Jones, Ove Arup, John Laing, Li Ka Shing, Bill Gates). Such people, through influence over employment of staff, shape the values and behaviour of members of the organization to develop the organization's identity, both internally and externally. Thus, organizational cultures (and climates) are self-perpetuating: persons who 'fit' are hired and they 'fit' because they are hired; errors of 'fit' are subject to resignation or dismissal. Thus, organizational cultures develop through the necessity of maintaining effective and efficient working relationships amongst organizational members and stakeholders (both temporary and permanent).

Commonly, pressure for cultural change arises from external parties, particularly in situations of environmental turbulence and attempts to enter new markets. That is magnified in situations of organizational take-overs and amalgamations, which usually act as major perturbations in organizational culture and developments: 'Usually the corporate culture of the most powerful or economically successful company dominates' (Furnham, 1997).

Cameron and Quinn (1999) employ a 'competing values' model in which 'flexibility and discretion' are juxtaposed to 'stability and control' on one dimension; the other dimension juxtaposes 'internal focus and integration' and 'external focus and differentiation'. The resultant model (see Figure 3.4) yields four quadrants, each denoting a type of organizational culture:

- Clan
  Some basic assumptions in a clan culture are that the environment can be best managed through teamwork and employee development, customers are best thought of as partners, the organization is in the business of developing a humane work environment, and the major task of management is to empower employees and facilitate their participation, commitment, and loyalty.

  (ibid.: 37)

- Adhocracy
  A major goal of an adhocracy is to foster adaptability, flexibility, and creativity where uncertainty, ambiguity and/or information-overload are typical. Effective leadership is visionary, innovative and risk-orientated. The emphasis is on being at the leading edge of new knowledge, products, and/or services. Readiness for change and meeting new challenges are important.

  (ibid.: 38–39)

- Market
  The major focus of markets is to conduct transactions with other constituencies to create competitive advantage. Profitability, bottom line results, strength in market niches, stretch targets, and secure customer bases are primary objectives for the organization. Not surprisingly, the core values that dominate market type organizations are competitiveness and productivity.

  (ibid.: 35)

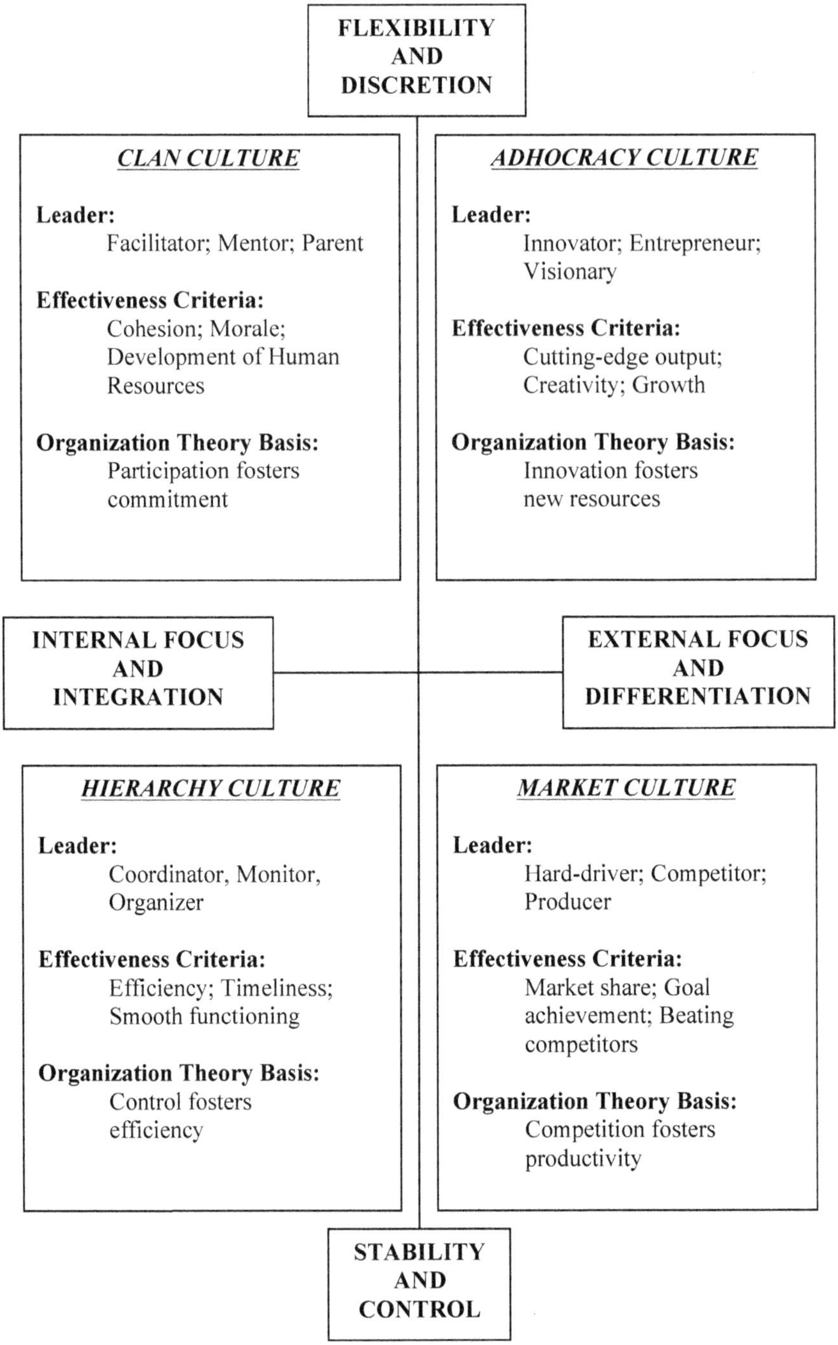

FLEXIBILITY AND DISCRETION	
CLAN CULTURE **Leader:** Facilitator; Mentor; Parent **Effectiveness Criteria:** Cohesion; Morale; Development of Human Resources **Organization Theory Basis:** Participation fosters commitment	**ADHOCRACY CULTURE** **Leader:** Innovator; Entrepreneur; Visionary **Effectiveness Criteria:** Cutting-edge output; Creativity; Growth **Organization Theory Basis:** Innovation fosters new resources
INTERNAL FOCUS AND INTEGRATION	**EXTERNAL FOCUS AND DIFFERENTIATION**
HIERARCHY CULTURE **Leader:** Coordinator, Monitor, Organizer **Effectiveness Criteria:** Efficiency; Timeliness; Smooth functioning **Organization Theory Basis:** Control fosters efficiency	**MARKET CULTURE** **Leader:** Hard-driver; Competitor; Producer **Effectiveness Criteria:** Market share; Goal achievement; Beating competitors **Organization Theory Basis:** Competition fosters productivity
STABILITY AND CONTROL	

Figure 3.4 Competing values and organizational cultures model (Fellows, 2006b: 57, following Cameron and Quinn, 1999).

- Hierarchy
  The organizational culture compatible with this form is characterized by a formalized and structured place to work. Procedures govern what people do. Effective leaders are good coordinators and organizers. Maintaining a smooth-running organization is important. The long-term concerns of the organization are stability, predictability, and efficiency. Formal rules and policy hold the organization together.

  (ibid.: 34)

The Denison model (Denison, 1997, 2009) also focuses on a set of tensions or contradictions. For example, the trade-off between stability and flexibility and the trade-off between internal and external focus are the basic dimensions beneath the organizational culture framework. The resultant quadrants comprise mission, consistency, involvement and adaptability, each comprising three constituents, as depicted in Figure 3.5. At the centre of this model there are underlying beliefs and assumptions. This is in recognition of the fact that the 'deeper' levels of organizational culture (Schein, 2004) are difficult to measure but provide the foundation from which behaviour and action spring. Denison employs a model comprising the same dimensions and quadrants to analyse leadership.

Schein (1984) determines two main types of organizational culture: 'free flowing' – an unbounded, egalitarian organization without a great deal of

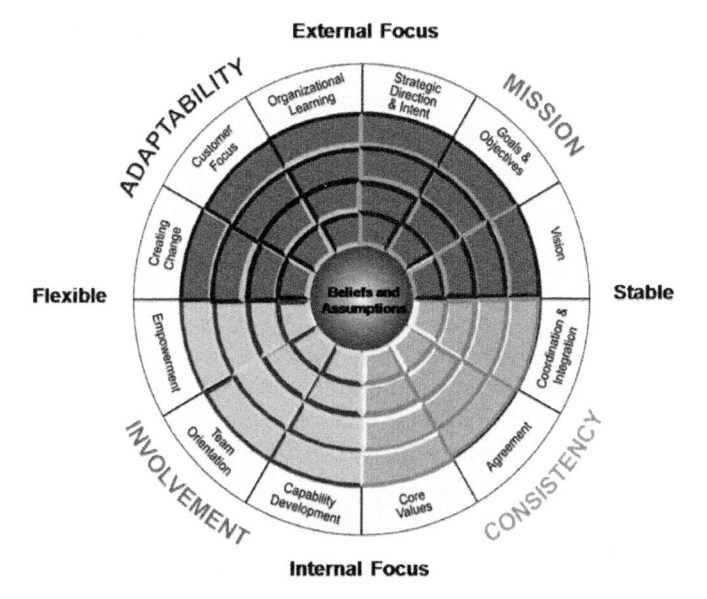

Figure 3.5 Denison's model of organizational culture (and leadership) (http://www.denisonconsulting.com/advantage/researchModel/model.aspx, accessed 7 October 2009).

formal structure, which encourages debate and internal competition; and 'structured' – a bounded, rigid organization with clear rules and requirements. (That categorization resembles the organic–mechanistic typology of Burns and Stalker, 1961.)

The dichotomy between formal and informal structuring is important in analysis of the realization of construction projects: formal systems are set up (organization charts, contractual procedures, etc.) but tend to be used 'only in the last resort' – when things go wrong. Projects operate through networks of informal relationships which emphasize 'doing what is practical/ pragmatic' to foster progress. The widespread concern is that strict adherence to the formal system would cause the project to 'grind to a halt' because of many bottlenecks (as in contract procedures regarding oral variations). However, the risks involved must be understood and accepted (which indicates that there is a low level of risk aversion in the industry). However, risk aversion should not be confused with Hofstede's (national culture) dimension of uncertainty avoidance (UA), which relates to willingness to try new things. Given the work allocation processes and (consequent) low profitability, coupled with performance pressure on time and quality, there is some reluctance to innovate unless the consequences are demonstrated to be unarguably beneficial; the low profitability is part of the rationale for low investment in research and development (R&D) in the industry. However, the other side of that same argument is that constructors on site innovate very frequently to overcome problems and secure or improve performance. Thus, it seems that UA in the industry is very context-dependent.

Handy (1985) suggests four types of organizational culture. A *power culture* is configured as a web with the primary power at the centre; emphasis is on control over both subordinates and external factors (suppliers etc. and nature). A *role culture* involves functions or professions which provide support to the overarching top management; emphasis is on rules, hierarchy and status through legality, legitimacy and responsibility (as in contractual rights, duties and recourse). In a *task culture*, jobs or projects are the primary focus, treating an organization as a net (as in a matrix organization); structures, functions, and activities are evaluated in terms of their contributions to achieving the organization's objectives. A *person culture* is one in which people interact and cluster quite freely; emphasis is on meeting the needs of members of the organization through consensus. Handy suggests that the main factors influencing which organizational culture develops are goals and objectives, history and ownership, size, technology, and environment and people.

Williams, Dobson and Walters (1989) employ the categories of 'Power', 'Role', 'Task', and 'People'. The categories (sequentially) correspond to those noted by Handy (1985).

Examination of the various, alternative sets of dimensions used to analyse national cultures and organizational cultures indicates considerable conceptual agreement. Generally, the dimensions used to analyse organizational cultures align with the task and human relations schools of management thought (such as McGregor, 1960 – Theory X and Theory Y; and subsequent development by Ouchi, 1982 – Theory Z).

Organizational climate

Tagiuri and Litwin (1968: 27) note that:

> Organizational Climate is a relatively enduring quality of the internal environment of an organization that (a) is experienced by its members, (b) influences their behaviour, and (c) can be described in terms of the values of a particular set of characteristics (or attributes) of the organization.

Hence, the climate of an organization distinguishes it from other, similar organizations. As organizational climate both reflects and shapes the work experiences shared by members of an organization, it indicates their perceptions about autonomy, trust, cohesion, fairness, recognition, support and innovation, and so leads to the members of the organization sharing knowledge and meanings. Organizations' climates are important contributors to relative homogeneity amongst members through recruitment and retention.

Organizational climate is linked closely with organizational culture. Organizational climate is rather less deep seated than organizational culture as it is grounded in the practices of an organization as experienced by its members (see Figure 1.1).

An important caution is emphasized by Moran and Volkwein (1992: 43) concerning desires to effect changes in organizational climate as a cognitive tool of management. There are

> implications for managerial practice in two respects. First, . . . since climate operates at a more accessible level than culture, it is the more malleable and, hence, the more appropriate level at which to target short-term interventions aimed at producing positive organizational change. . . . Second, the conceptualization . . . suggests that interventions to change climate must consider the deeper patterns embedded in an organization's culture.

Thus, change initiatives may not produce the desired effects.

Differences and changes

Cultural differences

Most problems are likely to occur across the boundaries of cultures as persons from different cultures interface with each other. The problems may be more frequent and significant if the differences between the cultures are not anticipated (such as can occur in discussion between engineers, architects and contractors from the same locality). If differences between people are obvious (appearance, language), allowances are much more likely to be made which increase tolerance; also more efforts are likely to be made to ensure that understanding is correct (by requesting clarification to check meanings) and to accommodate behavioural differences.

Cultural differences are potently manifested in languages – most obviously at national levels (the UK and USA are 'two nations separated by a common language'). Western languages tend to be high content (low context), in which each word has a precise meaning and words are assembled to convey exact messages. Eastern languages tend to be high context, in which the meanings of words are often vague (each word may have several meanings) but must be determined by the context of use in the message; also, it is common for messages to be constructed so as to be inexact and require interpretation through knowledge of the prevailing culture to reveal the meaning. Further, in Eastern societies, certain relationships, especially the initial phases of business contacts and those involving difficult decisions, may be conducted via an intermediary person who is known and acceptable to the primary parties; that serves to avoid direct conflict between the parties but does raise concerns regarding 'agency'.

Age is respected in most societies but, in employment situations, significant, culturally determined differences arise. In Western organizations, promotion is secured through achievement: educational or training qualifications and performance of work activities. In Eastern societies, the tradition is for promotion to be through ascription; employment is for (working) life and promotion depends on age and experience. In recent years, both 'traditions' are subject to some breakdown: in Eastern societies, redundancies are being invoked and more recognition is given to achievements; in Western societies, continuity of employment is recognized as valuable and permanence of relationships, together with stability and wisdom of age, is more valued.

There are two polar approaches to cultural awareness: one focuses on similarities and the other focuses on differences. Knowledge of both similarities and differences is important for understanding, appreciation and adaptation of behaviour – that is, to achieve compatibility – but there are potential negative aspects too. A strong focus on similarities may induce complacency and diminish sensitivity to important differences whereas a focus on differences may encourage incompatibility through ethno-centrism, 'jokes' about differences and so on, thereby inducing alienation.

The negative consequences of focusing on differences seem to be rather common in construction; this leads to participants' behaviour generating a culture of mistrust and disrespect in interpersonal or interorganizational relationships, which compounds the other negative perspectives which abound. To overcome such problems, change initiatives must be robust and high in valence of expected benefits with demonstrable, good prospects of success. Quite commonly, periodically shifting market forces prompt redistributions of power which lead to (temporary) changes.

Cultures and cultural changes impact on performance, including what aspects of performance are regarded as important, and so what forecasts are made and form the basis for selection of procurement approach and participants. Importation of requirements, methods, resources or personnel which are alien to the 'domestic' culture usually generates rapid rejection – whether overt (such as refusal to comply) or implicit (such as in low levels of performance). Analogous problems are encountered by many innovations. That situation is complex as many (sub)cultures exist contemporaneously.

Cameron and Quinn (1999) determine that organizations may not have a single, unitary culture and that metrics and measurements used to determine strengths of cultures are subject to significant question. However, Schneider (2000) believes that every successful organization has a core culture (control; collaboration; competence; cultivation) which is central to its functioning. Zhang (2004) studies the cultures and effectiveness of construction organizations in mainland China, and determines that a strong and balanced cultural profile (according to the organizational culture assessment instrument – OCAI; Cameron and Quinn, 1999) is most effective.

Those studies' findings are not necessarily incompatible. Many research findings, including those of Zhang (2004), Cameron and Quinn (1999) and Denison (2009), determine cultural profiles of organizations which include elements drawn from several dimensions; usually, the profile of an organization comprises a (more or less) dominant culture type but with other types present (some of which support but others countermand the dominant type).

Cultural distances

What cultural distances are and how they should be measured are issues of debate. Usually, cultural distance between two countries is measured by use of an index (as in Kogut and Singh [1988], who use Hofstede's [1980] initial four dimensions of national culture).

$$CD_j = \sum_{i=1}^{4} \left\{ \left(I_{ij} - I_{id} \right)^2 / V_i \right\} / 4$$

where i = a dimension of culture; j = the overseas country; d = the home country; CD_j = cultural distance of the overseas country from the home country; V_i = variance of cultural dimension i.

That approach to measuring cultural distances involves the issues of constructing and interpreting an index, including making assumptions which may not always be appropriate. The problems include that the index may be too simplistic to reflect the elements and subtleties of the culture(s) involved.

Measures of cultural distance as aggregate indexes of measures on dimensions of culture are challenged through concerns over the relative sizes of in-group and between-group variances. Further, cultures are dynamic temporally and vary within national borders. Not all cultural facets are of equal importance, nor do they necessarily operate in the same direction. Intra-cultural variations (national and organizational) may exceed inter-cultural variations (Au, 2000). Hofstede (1989) confirms that differences between cultures vary in significance and that differences in uncertainty avoidance are, potentially, the most problematic for international business alliances. However, some emphasize power distance, and others focus on individualism (vs collectivism – in its developing context of horizontal and vertical components).

For alliances, home country culture is embedded in the firm, whereas host country culture is embedded both in the alliance partner(s) and in the local, operating environment (Shenkar, 2001). Such an alliance is often organized to overcome perceived problems of organizational fit of the home firm into the host country (Jemison and Sitkin, 1986). By assigning responsibilities for operational management (labour, local suppliers, local institutions and government) to the host country partner, the potential detrimental effects (risks and costs) of cultural distance are reduced.

Cultural changes

Changes in organizational cultures occur, most usually, by gradual evolution in path-dependent directions which are likely to be punctuated by occasional periods of stability and other periods of rapid, step-type changes: 'The evolution of culture is shaped by agency and power, but cannot be created by fiat' (Weeks and Gulunic, 2003). However, 'despite agreement that cultural evolution occurs . . . , espoused approaches to culture interventions are more commonly revolutionary in nature' (Harris and Ogbonna, 2002). A desire to effect particular changes in the culture of the industry and its constituent organizations is evident in many attempted initiatives in construction, such as the introduction of the 'New Engineering Contract' and 'Partnering' (consider the expressed, desired effects of the '40-hour partnering workshop'). When faced with change, most people show a strong preference for the familiar and so tend to resist. Hence, even if change does occur, there is a strong tendency to revert to prior norms and procedures – the status quo (as in private finance initiative [PFI] projects – see, for example, Ezulike, Perry and Hawwash, 1997).

Perspectives on changes in cultures fall into two extreme categories. 'Functionalists' believe that organizational culture is a construct which can be subjected to direct control by management; hence, functionalists have been vociferous in promoting the notion of the cultural basis for determining organizational performance. The alternative perspective is to regard culture as a context within which action must be taken, so necessitating that actions be compatible with the cultural environment. A third category, which is an amalgam, is the perspective that culture may be malleable and so may be adapted under certain conditions – albeit that adaptations are likely to be very difficult, replete with ethical issues and problems, and require effort over long periods of time (as in the legal requirement, reinforced with prolonged publicity, to wear seat-belts in cars).

Cameron and Quinn (1999) employ their survey instrument not only to determine the organizational profile which employees perceive to exist in an organization but also to determine the profile of organizational culture which those employees prefer (the culture they would like the organization to have in five years' time). That is a valuable indicator to the management of what changes to the culture could be effected most readily but, of course, subject to the environment of the organization and the desires of other member groups and stakeholders.

It is important to note that it is probable that even the most carefully devised and conducted change initiatives will have unanticipated consequences – including ritualization of change, cultural erosion, hijacking of the process, and uncontrolled and uncoordinated efforts (Harris and Ogbonna, 2002). That indicates that such initiatives must be considered thoroughly and then their effects monitored and the mechanisms adjusted as implementation proceeds.

Awareness, understanding and accommodation of culturally based differences are recognized as important for successful performance because success is judged by culturally determined measures regarding technical and business performance, especially. Relational aspects are, increasingly, regarded as being of importance too. However, given the global impact of free-market (capitalist) economies and their performance imperatives (growth and profitability), both technical and relational performance are subjugated to financial performance.

Construction culture

Descriptions of the culture of the construction industry seek to capture its essential and unique characteristics; unfortunately, many such descriptions accentuate aspects of the industry which are problematic. Common cultural descriptors include 'low technology', 'macho', 'opportunistic' and 'claims oriented', and 'blame'. Low technology reflects the relative level of technology used in the industry's processes and in the output; despite changes – intelligent or responsive buildings and components (e.g. glazing) – organizations in the industry remain inhibited through risk levels, profitability concerns, and lack of investment in R&D. Macho reflects the traditionally dangerous, dirty, and physically demanding (of strength) nature of construction activities, which promotes the 'male domain' of being strong, tough, brave and so on to overcome the difficulties, employing loud voices, 'particular language' and muscular power – hardly essential now for the operations on highly effective and efficient (often Japanese) project sites!

Opportunistic and claims oriented reflect the processes and ploys which are adopted on occasions to win work, to enhance revenue and to reduce costs (hence, to increase profits) (see Rooke, Seymour and Fellows, 2003, 2004). However, certain of those activities can be seen from a different perspective: claims under a contract should constitute legitimate recourse (recompense) for defaulting or opportunistic behaviour of the other party. Anecdotally, in some countries it is very difficult to persuade local employees to submit and pursue contractual entitlements as claims because such behaviour is totally alien to their (local) culture and, hence, means of doing good (harmonious) business, particularly through preserving good relationships. Further, the opportunism and claims orientation seems to be, largely, caused by work allocation processes' orientation to price competition in respect of initial contract sum – the 'lowest bid wins' criterion.

An over-riding concern is to produce successful projects. That involves defining success; such a definition is likely to involve measures of performance, made at particular time(s), from specific participant(s)' perspective(s). A widely articulated objective is to 'satisfy the client'. Usually, the client is identified as the party who commissions the project (the 'immediate paymaster'). Alvesson (1994) notes a tendency amongst the professionals in advertising agencies to 'downgrade' the judgements and abilities of clients. A similar tendency is apparent also amongst design professionals and consultants on construction projects (see, for example, Fellows, Liu and Storey, 2004, 2009). That approach to clients' inputs occurs in spite of the necessity for clients and professionals to work together closely to achieve and enhance good performance.

Given a joint-venturing, mutual interdependence relationship between project participants, and the consequent performance, it is more appropriate to consider performance which yields (reasonable) satisfaction to all participants as constituting a successful project – a satisficing solution (Simon, 1956). Indeed, it is likely that the scope of those whose interests must be considered in determining whether a construction project is successful extends beyond the project participants to include the project stakeholders; from a sustainability perspective, that encompasses future generations.

The construction industry has known of its extensive communication problems for a long time (e.g. Higgin and Jessop, 1963) but little seems to have been achieved towards their resolution. Dismissive statements which abrogate responsibility to act ('it's the culture of the industry') remain common. People respond to communications in relation to their contents and contexts, the presence and importance of each of those elements depending on the nature of the society and the language used (the cultural context). Sensitivity to relevant others influences communications, interpretations of meaning and consequent action within a power hierarchical context (a construction project temporary multi-organization) of who those relevant others are (in a business context, at least).

Often, construction is said to operate in a highly volatile and turbulent environment (as discussed in Chapter 2); bidding may be somewhat random on account of impacts of extraneous factors and of inter-organizational manipulations (some of which may not be legitimate), but price competition alone is tending to decline in importance for work allocation. The importance of inter-personal and inter-organizational compatibility (Nicolini, 2002; Dainty, Bryman, Price, Greasley, Soetanto and King, 2005; Baiden, Price and Dainty, 2006) for achieving good performance, and hence project success, is becoming recognized.

Fragmentation is one of the most common and enduring criticisms of the construction industry. Alternatively, fragmentation can be described as specialization or division of labour which is considered conducive to improving performance. Fragmentation means that, throughout the realization of a project, there are distinct, transient power hierarchies. The various, often sequential/iterative, inputs required from independent, individual specialist organizations facilitate fragmentation. Following Lawrence and Lorsch (1967), fragmentation is problematic, not because of the functional separation or diversity, but because of the inadequacy of communication, coordination and integration – all of which are major cultural concerns. Usually, coordination is problematic so that real and extensive cooperation is quite rare.

Thus, given the extent of the divisory aspects of fragmentation coupled with the transient involvement of many parties to construction projects and the turbulence of the operating environment, many activities do not progress as planned. A notable consequential feature of the industry is that there are always many others available to blame for faults and room for manoeuvre is extensive. Thus, claims are widespread and frequently constitute a significant profit-earning opportunity in competitive pricing (see Rooke, Seymour and Fellows, 2003), disputes are common, overt innovation is low, and performance suffers.

Alliances

Introduction

It is evident that a paradigm shift is occurring concerning processes employed for the realization of construction projects. As has been noted above, the structure of the industry in many countries has changed significantly as 'main contractors' have ceased to execute construction operations themselves but, instead, become managers of subcontractors and suppliers – in effect all (such) projects are management contracts (in operational reality, if not under that form of contractual arrangement). Additionally, 'partnering' is increasingly widespread, if only formally or nominally (rather than by real adoption and practice of the philosophy) – and, for public sector clients, concession (consortia) arrangements of various types, notably PPP and PFI, are common.

Although much of the rhetoric for the use of concession arrangements, especially the 'overlay' of partnering supplements, concerns the alleged performance enhancements of procuring construction projects through such arrangements, a significant reality is the changes in the financing of those projects. Notably, for the (commissioning) client, using concession arrangements (with or without a partnership supplement) means that the project in use is secured by revenue expenditure (the completed project – prison, school, hospital premises – is rented from the private sector owner or concessionaire) rather than by a capital investment; for other projects – roads and the like – the users pay for the facility directly to the concessionaire through tolls, thereby avoiding any expenditure by the public agency for provision of the facility. Usually, such public–private concessions require ownership of the project to move (revert?) to the public sector (project promoter) authority at the end of the specified concession period (at which time, the condition of the project must be at the minimum specified in the concession contract).

In addition to the major changes in procurement, a variety of recently developed management paradigms are being applied to construction in various ways, including 'agile', 'lean', 'supply (or value) chains/networks', 'value management' and 'whole-life management'. A common, primary focus of those paradigms is on the customer, most often identified as the commissioning client, and the desire to 'satisfy the customer' through the provision of customer-oriented (determined) value. However, the chaining and network perspectives, in particular, are testament to the importance of the values of all the project members and stakeholders, the interdependencies between them, and their resultant impacts on project realization outcomes – manifested in the efficiency of project realization processes and the effectiveness of the realized projects in use.

The value accumulation mechanisms and pathways which underpin project realizations, irrespective of the formal procurement process adopted, demonstrate the inevitable and universal joint venture nature of construction projects. However, following analyses of formal alliances, Das and Teng (1999) note that, 'Because of incompatible organizational routines and cultures, partner firms often do not work together efficiently'. In the extreme, failure results. The failure rate amongst alliances may be even more than for 'start-up companies'; the failure rate of strategic alliances has been projected to be over 60 per cent (Anderson Consulting, 1999). Alliance Management International Ltd (1999) attributes 50 per cent of alliance failures to poor strategy and 50 per cent to poor management.

A business alliance is 'an ongoing, formal business relationship between two or more independent organizations to achieve common goals' (Sheth and Parvatiyar, 1992). Within the numerous forms of long term, inter-firm associations (coalitions, joint ventures, licences, supply agreements, etc.), there are 'two basic organizational modes of alliance: equity joint ventures (EJVs) and non-equity joint ventures (NEJVs)' (Glaister, Husan and Buckley, 1998). That classification relates to formal alliances but informal alliances are much more frequent; therefore they cannot be equity alliances but often constitute a hybrid of contractual alliances in which the contract binds the participants formally, while the wider, operating alliance is determined through informal, usually inter-personal, relationships.

Contractual alliances provide much greater entry and exit flexibility for participants and at much lower cost. However, such apparent advantages result in reluctance of the participants to make significant alliance-specific investment (Pangakar and Klein, 2001); the alliance partners undertake lower levels of investment in the alliance (less asset specificity, thereby reducing costs of potential withdrawal). In equity alliances, the partners' investments in the alliance promote the alignment of objectives and performance incentives which act to deter participants from both free-riding and opportunistic behaviour (ibid.).

For alliance or joint venture formation and operation, two relational factors are important. Bridging concerns the external linkages between individuals and/or institutions which assist in defining their relationships, and consequently their operations and performance. Bonding, analogously, focuses on internal linkages. Bridging and bonding may proceed interactively and are culturally determined – especially with respect to individualism / collectivism.

Trust

Uncertainty and trust are the two primary constructs which affect relationships and their institutional arrangements in formal alliances (Sheth and Parvatiyar, 1992). Bachmann (2001) views trust and power as means for social control within business relationships. Those concerns are commonly manifested in the criteria for selection of partners and the establishment of safeguards against opportunistic behaviour by (other) alliance members, thereby increasing *ex ante* costs in the business (relationship) venture (Williamson, 1985). However, Baiden, Price and Dainty (2006) find that, in the usual processes employed for selecting the participants to realize construction projects, the vastly dominant criteria continue to be price and the (perceived) technical expertise/capability of the organizations; the ability of the organizations to integrate and cooperate to deliver the project effectively and efficiently remains ignored.

Trust, the antithetical complement of risk, is always an element in the decision to engage in a (business) relationship, whether the source(s) of trust is perceived to be in legal or contractual mechanisms, in institutions (Hagen and Choe, 1998; Bachmann, 2001), or in individuals (singly or in combination). Generally, trust is defined as 'Confidence in or reliance on some quality or attribute of a person or thing, or the truth of a statement' (OED, 2009) or, in relationships, it may be considered to be adequate confidence (on the part of the subject actor or participant) that the other participant(s) will not cheat – that the other will not behave to cause detriment to the actor.

Trust may be classified in several alternative ways, the most appropriate depending on the situation. For initial, or individual, encounters between actors, dispositional trust is appropriate (do the actors intuitively trust others or are they cautious, wary of others and so take precautions for their self-protection?). Additionally, trust may be based on the reputation of the other (reputational trust), which then operates to modify dispositional trust. For recurring encounters, trust may be based on experience (of others generally and of the particular others) – experiential trust. Trust may relate to organizations (and through their agents) – inter-organizationally and between organizations and individuals (see, for example, Lau and Rowlinson, 2009).

Yamagishi and Yamagishi (1994) distinguish trust and (performance) assurance in that, if one actor trusts another because the former has some provisions in place so that the latter has an incentive to cooperate, then that situation is one of assurance; trust exists when an actor believes that the other has an incentive to 'cheat' but does not do so – perhaps because of goodwill.

Hagen and Choe (1998) note that 'a trust relationship in business involves an expectation of cooperation but not an expectation of altruism'. 'The Chinese system of networked transactions . . . is relatively uncodified, and it is based on trust and long standing personal connections' (Boisot and Child, 1996). Thus, social institutions give rise to differing levels of trusting behaviour through the required behavioural norms (and limits) and sanctions for transgressors (who are caught); those institutions also impact on trustworthiness of individuals and organizations and, hence, on apparent dispositional trusting behaviour.

Thus, 'The relationship . . . in Japan is *not* built primarily on trust, but on the mutual interdependence enshrined in the agreed-upon rules of the game' (Womack, Jones and Roos, 1990: 155, emphasis added). That is afforded more general applicability by Hagen and Choe (1998), who state that 'the institutionalized industry practices that we call "institutional sanctions" in the context of societal sanctions are key determinants of interfirm cooperation', as manifested as the deterrent-based trust in Japanese industry.

Although trust and distrust are commonly viewed as opposite ends of a single dimension, that may not be appropriate (Lewicki, McAllister and Bies, 1998). Given that many business relations are multi-faceted, 'relationship partners might trust each other in certain aspects, not trust each other in other respects, and even distrust each other at times' (ibid.). That finding suggests the contingent nature of relationships together with the importance of culture and perceptions of cultures in determining behaviour.

Control

Risks in business alliances may be classified as relational risks and performance risks (Das and Teng, 1999). Relational risk, the risk of unsatisfactory inter-firm cooperation, includes lack of commitment by partner firms and their possible opportunistic behaviour; performance risks are all other risks (environmental, market and internal). Thus, participants endeavour to protect themselves from the risks they perceive by effecting control through contracts, equity and managerial means.

Equity control is a very effective deterrent of opportunistic behaviour by alliance partners as the ownership stakes reflect and promote alignment of partners' interests; however, such arrangements are, by definition, applicable to formal alliances only and give rise to significant costs of exit (Steensma, Marino, Weaver and Dickson, 2000). For non-equity alliances, both contractual and managerial control are available. Managerial control is effected by ensuring that staff from one's own organization occupy posts within the alliance which are critical to the alliance's performance; regular meetings between participating organizations help. Further managerial control to enhance the performance of alliances may be achieved through aligning the organizational routines (administration procedures and processes) of the partners – more likely to occur in enduring rather than temporary relationships. Essentially, the issue is one of organizational fit – strategic, resource and procedural.

Thus, Teece (1986), adopting a resource-based view, determines that alliances depend on their ability to identify and to use valuable resources – those which are rare, difficult to copy and irreplaceable – for success. Considering that power arises from control over resources, any significant imbalance may lead to erosion of trust and its replacement with emphasis on power in the governance of relationships so that 'In situations of power imbalance there is a temptation to enforce cooperation through power rather than trust' (Korczynski, 2000).

Oliver (1997) confirms that the culture of an organization is strongly influenced by both the national culture(s) and other aspects of the institutional environment(s) in which it has operated. Thus, managers in different countries make different strategic decisions because they possess different cultural values (Hofstede, 1980; Schneider and DeMeyer, 1991). Shane (1994) argues that national differences in levels of trust impact on perceptions of transactions costs and, thereby, influence the desirability of internalization and the choice of entry mode to new (foreign) markets. The dominant values of a national culture are usually reflected in the organizational culture (Hellriegel, Slocum and Woodman, 1998). In the case of an international joint venture, both the national culture and the organizational culture of the dominant partner are likely to be primary determinants of the organizational culture of that alliance.

Kahneman and Lovallo (1993) discuss a variety of aspects of human behaviour with particular reference to decision-making issues. They determine that a common illusion is the extent of control which people believe they can exert – greatly in excess of the reality. They note that, commonly, 'Managers accept risks, in part, because they do not expect that they will have to bear them' (ibid.). A further factor is optimistic bias, which comprises unrealistically positive self-evaluations (especially concerning forecast accuracies) and unrealistic optimism about future events and the probabilities of plans being realized. Finally, control efforts diminish rapidly, if outcomes are perceived to be successful – that is particularly dangerous as success is, in many instances, more likely to be due to chance than to control.

Sustainability

Introduction

Commencing with increases in fuel costs due to the first 'oil crisis' of the 1970s, many people have become aware of the impacts of energy usage. At first, concerns related to costs of fuel to occupy and use buildings, often coupled with frustration at the paucity of forecasts produced by the models available. That fostered emphasis on data collection and modelling of occupation costs of buildings as a rapidly growing element of life-cycle costing exercises to aid effective design of premises for efficiency of insulation: balances between energy consumption and cost.

Widespread and growing concerns over fossil fuel depletion, and depletion of resources generally, together with sources of energy and problems of pollution and global warming, extend the issues into concerns for the environment and for sustainability.

If humans and other living species are to survive, changes must occur. Hence, questions concern what changes are required, what are the consequences of those changes, who is to implement them, and how. As the cause of the problems, it seems morally incumbent on humans to address them and to strive for their permanent resolution, irrespective of how unpopular or uncomfortable they may prove to be.

Human behaviour

Human behaviour, simplistically, is driven by two types of force: innate drivers for survival (genetic) and cognitive forces, including learned and intuitive behaviour. Therefore, much human behaviour is 'selfish' but much cognitive behaviour pays regard to others.

Models of behaviour (e.g. Hofstede, 1980, 2001; Schein, 2004) identify core underpinnings to be people's beliefs and values. The stronger a belief, the more difficult it is to change and the greater its influence on behaviour. Morals are beliefs of what is right and what is wrong, regarding both means and ends, and thus are powerful influencers of behaviour.

An important question is 'how are people motivated?' Usually, motivation begins with a prospective benefit which is valued by the motivatee, who perceives a (causal) positive, usually quantitatively progressive, link between behaviour and receipt of the benefit. To be effective, and to remain so over time, the system must be communicated and operated properly, with control, to ensure the appropriate linkage between perception, effort and outcome. A problem with motivators, especially money, is that their motivational effectiveness (impact on behaviour) often diminishes rapidly over time.

A particular problem for sustainability is that behaviour which is required to promote it involves costs for the individual but quite uncertain and unknown benefits, the incidences of which are unknown too and may be far distant in the future. Hence, the primary aspects of motivation – for direct and immediate personal gain – do not apply. Instead, different personality variables operate involving long-term and collectivist beliefs and values such that the envisaged gains are valued more highly than the immediate costs. Thus, behaviour towards sustainability is likely to be grounded in moral beliefs and values and to include altruistic behaviour.

Within studies of culture, certain dimensions have been determined which are germane to behaviour conducive to sustainability. Hofstede (1980, 2001) included dimensions of (individualism–)collectivism and (short-term–)long-term orientation; Trompenaars and Hampden-Turner's (2005) dimensions included (individualism–)communitarianism, human–nature relationship and human–time relationship. Culture exerts extensive influence over individuals' behaviour and so the more collective and long-term is the cultural orientation, the more likely members of such societies are to behave in ways conducive to sustainability. Such behaviour is likely to be enhanced if people believe that they are not in great control (over others or over nature) and so are subject to what occurs in the environment; that tends to induce such people to endeavour to live in harmony with their environment, thereby protecting and sustaining it.

Sustainability and 'greening'

The OED defines sustainable as 'Capable of being borne or endured; supportable, bearable ... Capable of being maintained at a certain rate or level.' To endure is *to last*. Hence, something is sustainable if it continues to exist – at the limit, for ever.

Generally, definitions of sustainability follow that of the World Commission on Environment and Development (1987) (the Brundtland Report): 'development that meets the needs of the present without compromising the ability of future generations to meet their own needs'. Countries participating in the 1992 United Nations Conference on Environment and Development (The Earth Summit) in Rio de Janeiro agreed an action plan for the twenty-first century: Agenda 21. That plan recognizes that humans depend on the Earth to sustain life and that there are extensive and inexorable linkages between human activity and environmental consequences. Chapter 7 of Agenda 21 specifies the overall objective of human settlement to be 'to improve the social, economic and environmental quality of human settlements' (Agenda 21, 1992). Such statements propose notions of different *forms* of sustainability.

The UK government suggests that the principles of sustainable development comprise (DETR, 1999):

- maintaining high and stable levels of economic growth and employment;
- prudent use of natural resources;
- effective protection of the environment;
- social progress that meets the needs of everyone.

Those principles have been developed into principles for sustainable construction (DETR, 2001):

- constructing projects that are more cost-effective to produce and run as they have been constructed with less and yield more;
- constructing projects that contribute positively to the surrounding environment, using materials and systems that are easily replenished and perform better over their full life cycle;
- promoting high standards of living for people.

For sustainability, the resources of the Earth can be used only up to the rate at which they are replenished. Today, such consumption of resources is being far exceeded. 'The current economic growth in developed countries entails high rates of consumption of natural resources that nature is unable to restore, and great amounts of residues that cannot be absorbed' (González-Benito and González-Benito, 2005). That situation is applicable to all countries, not only 'developed' ones. What we are doing is using up many of the world's natural resources, transforming them into both desired and undesired (waste, pollution) forms and failing to replenish the resources in either the original or sufficiently close substitute forms.

Science indicates that the world comprises matter and energy as fundamental, interchangeable constituents ($E = mc^2$). Clearly, given global warming, our world is not a closed system and, with our current knowledge of 'black holes', is not truly sustainable; for more practical purposes, it is helpful to consider the matter and energy of the planet in terms of forms and quantities of each form, coupled with a perspective of changes over time. Further, it is appropriate to endeavour to identify influences on those quantities and changes and to determine what may lie within the control of humans.

A major problem is that, often, 'sustainability' is no more than a label used in discussion – what is really being debated is not sustainability but a related and much less demanding topic – 'greening'. Cole (1999) classifies 'green' performance of buildings as being assessed in relative/comparative terms, perhaps including benchmarking, but 'sustainability' performance assessments use absolute measurements (of energy embodiment, consumption, etc.). Although 'greening' is worthwhile, it is only a move towards the potential achievement of sustainability and, on many occasions, only a very small step!

Some evaluation issues

The only viable option is to adopt an absolute understanding of what sustainability is – relating to the total system of the planet, its environment and interplay with the cosmos via its permeable boundary. The effects of 'individual/subsidiary categories of sustainability' can be examined individually but a holistic analysis is essential to determine what is sustainable and what is not.

Because the absolute definition is so rigorous and perceived as highly restrictive (condemnatory of the actions of most people and societies), the issue is commonly 'fuzzed' through adopting a convenient-to-purpose, loose definition or no definition at all! Fuzzy definitions promote fuzzy thinking and lead to inappropriate actions.

The International Organization for Standardization (ISO) has produced the ISO 14000 family of standards through which organizations endeavour to manage the impacts of their activities on the environment.

> ISO 14001:2004 does not specify levels of environmental performance. If it specified levels of environmental performance, they would have to be specific to each business activity and this would require a specific EMS standard for each business . . . An EMS meeting the requirements of ISO 14001:2004 is a management tool enabling an organization of any size or type to: identify and control the environmental impact of its activities, products or services, and to improve its environmental performance continually, and to implement a systematic approach to setting environmental objectives and targets, to achieving these and to demonstrating that they have been achieved.
>
> (ISO, 2009)

Thus, ISO 14000 compliance does not, of itself, protect the environment; it is a framework. The standards regarding environmental impact are specified elsewhere.

> BREEAM (BRE Environmental Assessment Method) addresses wide-ranging environmental and sustainability issues and enables developers and designers to prove the environmental credentials of their buildings to planners and clients. It uses a straightforward scoring system that is transparent, easy to understand and supported by evidence-based research, has a positive influence on the design, construction and management of buildings, sets and maintains a robust technical standard with rigorous quality assurance and certification.
>
> (BREEAM, 2009)

BREEAM and other assessment systems for environmental impacts of buildings and their operation are incomplete. Embodied energy and other resources, production and related assessments are inexact as they employ norms of quantities. Even the best-scoring solutions are not (necessarily) sustainable according to rigorous definition. Therein lies a significant danger: by achieving a good score, with flattering descriptors, perhaps being benchmarked as a 'best solution', it is all too easy to be lulled into regarding the building as sustainable. It almost certainly is not! It may be a valuable step in the direction towards sustainability, it may be a (relatively) very 'green' building – which is to be applauded; the essential thing is to appreciate it for what it is.

Motivation for sustainability is fraught with problems. Sustainability behaviour is altruistic and may not yield any detectable benefit. Sustainability is the most global, complex and long-term of constructs, thereby often invoking a perception of helplessness, immateriality of individual actions and necessity to overcome major barriers (high costs) by those who want to pursue sustainability. That includes issues of sorting waste for recycling, of walking rather than using transport (including lifts and escalators) and many, many other behaviours. 'Why should I, when nobody else does?' Why should developing countries forgo immediate gains by being less environmentally detrimental when the world's most developed and richest countries act in such environmentally obnoxious ways? Notably, the USA has no intention of ratifying the 'Kyoto protocol'; however, despite such glaring selfishness by the nation which, commonly, preaches moral standards and behaviour requirements to others, by August 2009, 183 countries had ratified and signed.

While initiatives such as carbon trading act to cap and, over time, reduce greenhouse gas (GHG) emissions through progressive reduction of permitted levels of emissions, the trading of emission (carbon) credits, which are sold and bought by emitting organizations, may be viewed as a short-term palliative for the greater emitters. The trading is an element of the 'polluter pays' principle and, even if the financial cost of buying emission credits includes a significant premium, it still allows the additional pollution to occur; although pragmatic, carbon trading is a short-term and mild alternative to enforced pollution capping and progressive reduction (continuous improvement, as in the form of carbon offsets to reduce carbon footprints, the main subset of ecological footprints).

In cost–benefit analysis (CBA) terms, sustainability behaviour incurs large private costs to yield small private benefits but with possible large external benefits, especially in the longer term. The global nature of the issues requires collective, universal action. What some are pursuing today is, in practice, 'greening': a useful step towards sustainability but, still, only a step – 'every little helps'! Both complacency and helplessness are perceptions highly detrimental to sustainability endeavours.

There is a vast amount of political manoeuvring in the sustainability arena, commonly accompanied with much propaganda; only through accurate understanding of the concepts and issues can real sustainability be pursued.

Sustainability, incorporating resource consumption and pollution removal, is a global matter; indeed, only if the component problems are tackled globally can sustainability be possible. Humans suffer from delusions of control – we can control far less and to much more limited extents than we like to believe. Perhaps the major difficulty, however, is subjugating current self-costs for the essential, but altruistic, long-term benefit of others.

Conflict and disputes

Introduction

Conflict is 'the clashing or variance of opposed principles, statements, arguments, etc.', whereas a dispute is 'an occasion or instance of . . . an argumentative contention or debate, a controversy; also, in weakened sense, a difference of opinion; freq. with the added notion of vehemence, a heated contention, a quarrel' or 'strife, contest; a fight or struggle' (OED, 2009). According to Van de Vliert (1998), 'Two individuals, an individual and a group, or two groups, are said to be in conflict when and to the extent that at least one of the parties feels it is being obstructed or irritated by the other'. More popularly, conflict is 'incompatible behaviour between parties whose interests differ' (Deutsch, 1973).

Conflicts and disputes occur along a spectrum from a minor difference of opinion to full litigation with appeals up to the Supreme Court. Although 'conflict' is a generic term which covers the whole spectrum, in business and construction practice conflict is regarded as lesser and informal, a disagreement; a dispute is a conflict which has escalated and been formalized towards a law-determined outcome.

In businesses operating in market capitalist economic systems, competitive procedures and contexts (e.g. tendering for construction projects) coupled with performance imperatives for survival (notably, profitability) operate to set participants in opposition to each other in a zero-sum game of pursuit of self-interests. As such a situation is virtually ubiquitous in the world's construction industry, it is hardy surprising that the industry is perceived to have a 'culture of conflict', with high incidences of conflict and disputes and extensive opportunistic behaviour.

Often, conflict arises through frustration. A person perceives the actions or inactions of another(s) have a detrimental effect on the former's potential achievement– usually expressed in terms of reduction in extent of (anticipated) goal attainment (McKenna, 2000). Thus, conflict arises from, and is likely to magnify, negative feelings.

Especially in management 'teams', conflict is classified as cognitive (intellectual and technical issues) or affective (subjective and emotional aspects) (Amason, Hochwarter, Thompson and Harrison, 2000). That typology is reflected in much of the pragmatic conflict resolution literature, such as Fisher and Ury (1991), which strongly advocates focusing on the cognitive issues to achieve resolution on account of the negative, destructive and blocking potential of the affective aspects.

Conflict may be analysed and classified in terms of its effects: either functional/constructive or dysfunctional/destructive. Often, functional conflict is aligned with competition (parties have a common objective, even if the outcome is of win–lose form, a zero-sum game, as in construction bidding), as it is believed to stimulate ideas, innovation and so on, and thus operate as a motivator. Dysfunctional conflict, however, yields the well-known, undesirable, negative consequences which are detrimental to relationships and technical performance and, consequently, business performance.

There are three primary perspectives on conflict. The traditional, functionalist perspective views conflict negatively: as disruptive and, therefore, harmful to performance, people and their relationships. The behavioural perspective is neutral: conflict between individuals and groups is considered to be inevitable; differences between the consequences of conflict occur owing to differences between people – perceptions, personalities, interests/expertise and goals. The interactionist perspective views conflict positively: as carrying out useful functions to ensure social dynamism and to enhance decisions; however, the conflict should be managed, both the type and extent of conflict, so that its consequences remain positive and the conflict is not allowed to escalate to produce negative effects.

Sources of conflict tend to be classified as communication, structural and personal factors (Robbins, 1974). In addition to the well-known communications issues which may give rise to conflict – semantic differences, insufficient or excess information, noise, and filtering of information (distortion, withholding, etc.) – other aspects, including indexicality (Clegg, 1992), choice of channel(s) and the nature of the language (high content or high context), impact too. The structural factors comprise:

- size and constituent members of the group – the more diverse, the greater is the scope for conflict;
- ambiguity – positively correlated with conflict;
- leadership – participatory leadership may encourage conflict (by revealing differing views);
- rewards – win/lose system (competition; zero-sum game) is conducive to conflict;
- interdependence – emphasizing common goals (benefits) reduces conflict which, otherwise, may arise from diversity of expertise, perspectives and interests;
- changes to structure and/or processes – people tend to resist change, especially if they view the consequences as detrimental.

The personal factors include personality characteristics (traits etc.) and people's beliefs and value systems (fundamentals of culture).

Various models of the conflict process have been advanced: notably, the bargaining model, the bureaucratic model, the escalation (and de-escalation) model (van de Vliert, 1998) and the systems model, which develops into the episodic model (Pondy, 1967). The models recognize that conflict incidents (episodes) are not isolated, individual events but cycle iteratively, each episode having antecedents and enduring consequences which impact on people's dispositions through experiential learning and thus generate the antecedents of subsequent conflict episodes. The escalation/de-escalation perspective is determined by how an episode is managed and therefore yields the nature of an episode's consequences as antecedents of potential, subsequent episodes (see Figure 3.6).

Thus, a conflict episode reflects both the current subject matter of the conflict and the psychological aspects remaining from preceding conflict episodes, both processes and outcomes. Those psychological aspects (antecedents) combine with the conflict components of principles and positions regarding the instant conflict: the principles relating to the subject matter of the conflict (cognitive) and the positions relating to the emotions of the participants (affective).

The spectral nature of conflict episodes reflects a number of factors: duration, formality, detail and complexity, growth of affective aspects, increasing entrenchment of parties, costs, changes in basis of resolution (towards legalities) and amendment of focus (from resolving the technical issues to 'beating' the other party). A step-change in focus is virtually inevitable in the progression at the point when the conflict is formalized into a dispute – commonly, when correspondence quoting contractual terms commences.

Resolution

Resolution of conflicts (and disputes) may be effected by managerial, technical–legal and law-based methods. That classification reflects a major change in orientation from a blend of relationships, technical and business issues, through a mix of technical and business issues with an 'eye to the law', to law-determined methods based on the technical evidence presented, using highly formalized procedures (which significantly distance, even alienate, the disputants from the resolution process). Whereas managerial methods may be informal (quick and inexpensive) and private, formality, duration, expense and public exposure increase rapidly in moving towards a law-based outcome. Enforcement (for certainty of performance) of the outcome (solution) is an important consideration together with other consequences – notably, effect on (business) relationships – both particular, as between the participants (notably, trust and trustworthiness), and general, in respect of the participants' reputations (how others perceive them – notably, assurance via social institutions).

76 *Culture's influences in construction*

```
                    ┌─────────────┐
                    │ Antecedents │
                    └──────┬──────┘
                           │
              ┌────────────▼────────────┐
              │     Latent Conflict     │
              └────────────┬────────────┘
     ┌──────────┬──────────┼──────────┬──────────┐
┌────▼─────────┐                      ┌──────────▼───┐
│ Feelings and │◄────────────────────►│  Issues –    │
│ Perceptions  │                      │ Technical,etc│
│  [Affective] │                      │  [Cognitive] │
└──────────────┘                      └──────────────┘
                    ┌────────▼────────┐
                    │ Manifest Conflict│
                    └────────┬────────┘
                             │
          ┌──────────────────▼───────────────────┐
          │        Conflict Behaviour            │
          │    (escalative / de-escalative)      │
          │    (spontaneous / calculated)        │
          └──────────────────┬───────────────────┘
                             │
                    ┌────────▼────────┐
                    │  Consequences   │
                    │   / Aftermath   │
                    └────────┬────────┘
                             ┊
                    ┌────────▼────────┐
                    │     Future      │
                    │    Potential    │
                    │    Conflict     │
                    │   Episode(s)    │
                    └─────────────────┘
```

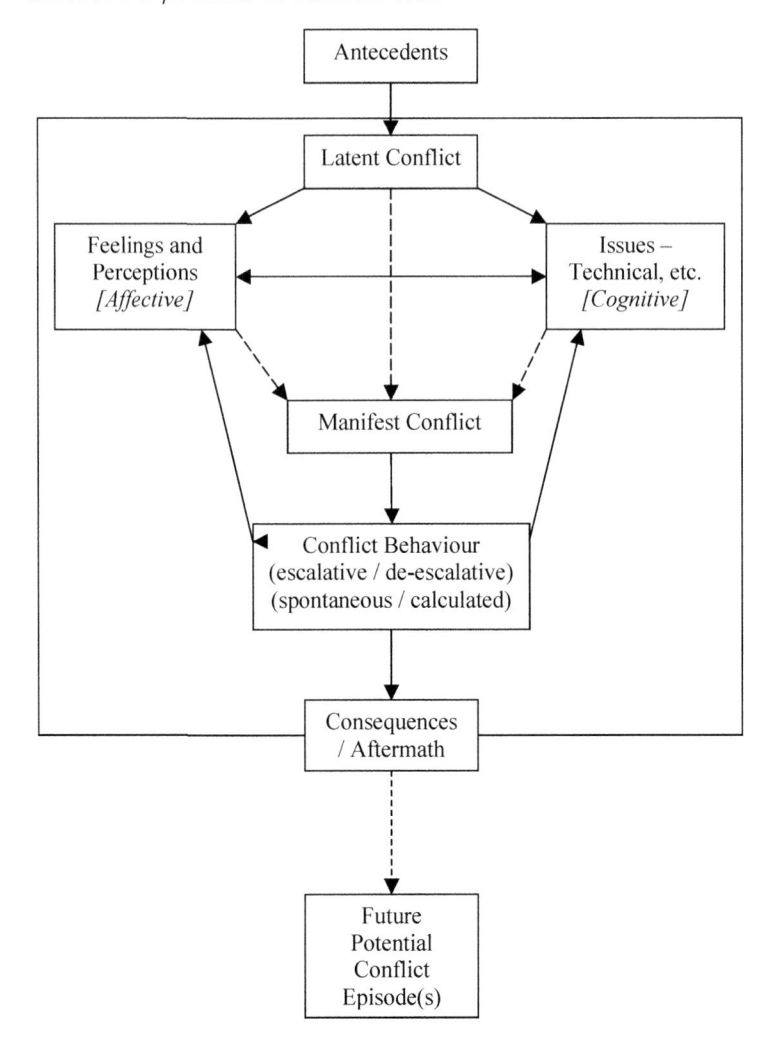

Figure 3.6 A conflict episode (adapted from Pondy, 1967, and Van de Vliert, 1998). Note: some conflict episodes may remain latent but will remain 'festering' and impact on future episodes as antecedents.

Selection of the resolution method is complicated and depends on a variety of variables, including the nature of the conflict or dispute (cognitive and affective aspects); the stage to which it has progressed; the prevailing conflict or dispute resolution processes available and legal contexts; the relative size of the conflict or dispute to the parties, each party's resources available for resolution and each party's relative power; the personalities of the parties; and the advice obtained by the parties. Choices of resolution methods, in most cases, are not mutually exclusive, although some may be imposed by prevailing law (e.g. in the UK construction disputes should be resolved through arbitration, rather than litigation), and may occur along the spectrum until resolution is achieved.

Examination and discussion of the formal resolution processes (quasi-legal and legal) is outside the scope of this text. Hence, the examination below concerns management of the resolution of conflict.

Conflict (resolution) management

Managerial methods for achieving resolutions of conflicts have a major focus on preventing escalation of conflicts. Given this, the methods are operated by the parties themselves, although external expertise may be used also (notably, mediation).

Generally, five categorical styles of conflict management (handling or coping) are considered: avoiding (unassertive and uncooperative), competing (assertive and uncooperative), collaborating (assertive and cooperative), accommodating (unassertive and cooperative) and compromising (mid-assertive and mid-cooperative). Those styles fill the two-dimensional space of conflict management between axes of assertiveness and cooperativeness (Thomas, 1992), as depicted in Figure 3.7.

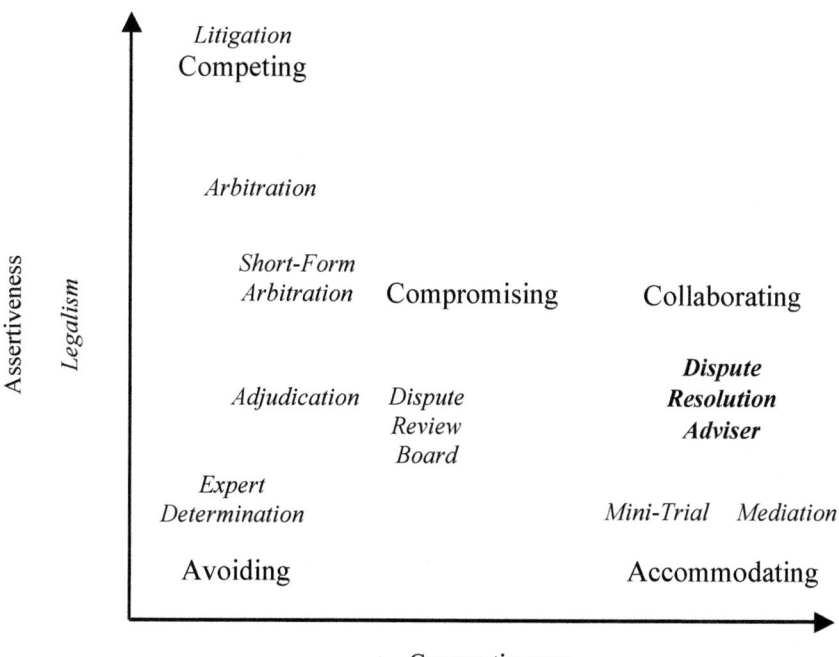

Figure 3.7 Styles of conflict management (following Thomas, 1992) and dispute resolution methods.

Avoiding involves suppression of the conflict and/or withdrawal (sometimes regarded as a separate style of conflict management) so that the subject matter is not addressed and so is likely to remain dormant or festering for others to resolve later. Competing takes the form of a zero-sum (win–lose) game in which one's own interests are pursued, often aggressively, by use of authority, power and the like. Collaborating seeks to pursue mutuality in seeking a solution to yield a non-zero-sum (win–win) game and to preserve or even enhance relationships through striving together for a solution. Accommodating resembles appeasement by subjugating one's own interests to those of the other(s), which tends to take the form of a zero-sum game. Compromising, again, seems to be a zero-sum game of give and take to yield a solution of balanced gains and losses for each.

Conflict behaviour is a person's outward reaction to the conflict which is felt or perceived, so the components of such behaviour are interrelated. 'Interpersonal conflicts really are complex situations in which different motives and concerns about own goals, the reaction with the other, other's goals, as well as short and long-term objectives, direct behaviour' (Euwema and van Emmerik, 2007). Thus, the conflict behaviour of the individuals involved determines the progression of the episode and the means of its resolution: whether a solution is negotiated by the parties or, following escalation to a dispute, a solution is imposed through operation of a (quasi-)legal system which alters the focus of resolution from the subject matter to the law (based on the evidence presented regarding the subject matter).

Sherif (1967) found that identifying and pursuing a superordinate goal(s) is effective in resolving inter-group conflict – the groups identify a common problem which dominates, and cooperation between them is required to achieve resolution. However, significant disparity in the distribution of power between the groups may preclude such resolution.

4 The practice
International case studies

Introduction

As an example of several different views and roles, incorporated into real-life construction projects, a series of documented case studies have been selected. The basis for the description of these case studies is a framework used as a structuring tool for discussing and analysing the issues raised therein. The cases studied are in general parts of larger total construction processes, thus representing snap-shots of critical incidents within the daily practice of the parties involved.

A framework for structuring and analysing the case studies: the 3C-Model™

An often used management tool for investigating (behavioural) experiences in construction projects and processes is the so-called '3C-Model™', as a structuring framework for analysing personal relationships within construction processes as examples of business processes. This framework was developed during an investigation and research within international real-life construction projects and processes in the period between circa 1992 and 1996 (Tijhuis, 1996). Additional experiences and analyses proved that the influence of personal relationships in such construction processes was and is still real, although the rise of, for example, the 'dot.coms' and 'social media' has influenced and will influence the character of such personal relationships (Tijhuis, 2001a). As also used in this chapter for the presented case studies, the following three aspects and their interconnections are analysed:

1 Contact – culture;
2 Contract – project organization;
3 Conflict – technology.

The three aspects Contact, Contract and Conflict together form the basis for the '3C-Model™'. The three aspects are connected by a framework, characterized as a kind of 'continuous loop'; this means that in fact this 'loop' represents a kind of possible repeating process, during one or more

80 *The practice: international case studies*

(possibly partial) iterations of which the several behavioural experiences can be structured and analysed within construction processes.

An overview of the used 3C-Model™ is represented schematically in Figure 4.1 (Tijhuis, 1996).

The three aspects *Contact*, *Contract* and *Conflict* are related on the the basis of 'levels of influence'. In this model, the 'drivers for change' of processes (Contract) within project organizations mainly proved to be 'culture' (Contact) and 'technology' (Conflict). For construction processes, this is represented schematically in Figure 4.2 (Tijhuis, 1996).

On the basis of this developed framework, the importance of the personal relationships in construction industry is highlighted, despite actual modern or post-modern developments related to, for example, the aforementioned 'dot. coms', 'social media' and the like. This viewpoint is also supported by several research outcomes from business culture-related research within, for example, the former CIB Task Group TG 23 (Fellows and Seymour, 2002) and its follow-up, the CIB Working Commission W112 'Culture in Construction'. As a result it was explained then that extra knowledge and insight of business culture issues is needed within an international context, as was, for example, indicated by a collaboration project of the authors, being partners from Hong Kong/China and the Netherlands (Liu, Fellows and Tijhuis, 2002).

The results of the case studies therefore try to add more insight-information and practice-experiences to this research field especially, related to different human behaviour in overall construction processes.

The selected case studies

The following case studies have been selected and are described and analysed using the described 3C-Model:

Case study 1: The Netherlands
Developing a complex inner-city project

Case study 2: Germany
Construction of rationalized serial housing

Case study 3: Poland
Subcontracting infrastructural and foundation works

Case study 4: Turkey
Tendering for developing a production factory

Case study 5: United Arab Emirates
Designing a production factory

Case study 6: China
Developing export markets for special building products

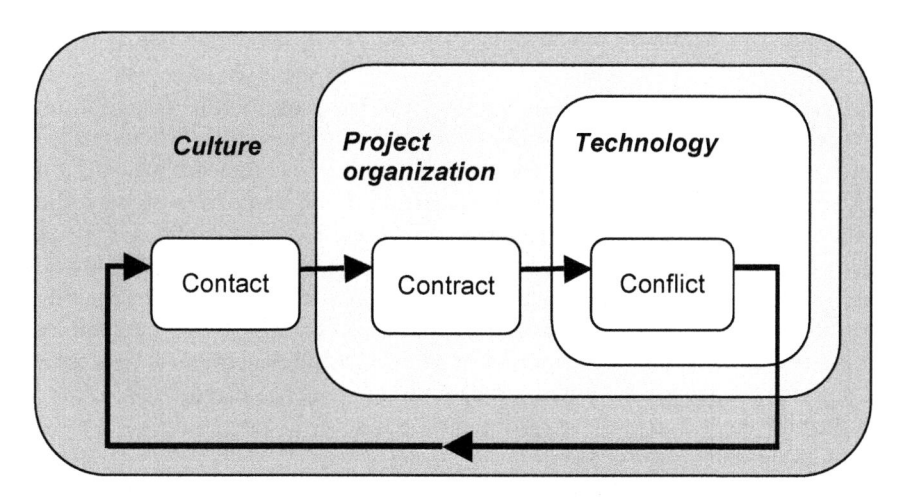

Figure 4.1 The 3C-Model™ as a framework for investigating (behavioural) experiences in construction processes (Tijhuis, 1996).

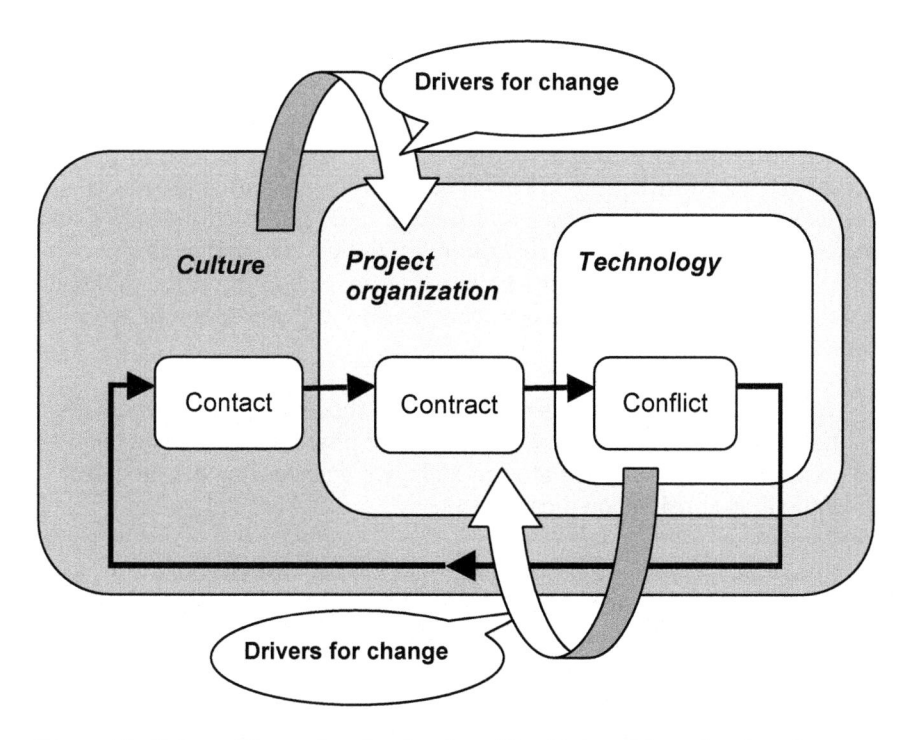

Figure 4.2 Culture (Contact) and technology (Conflict) as drivers for changing construction processes (Contract) within project organizations, based on the 3C-Model™ (Tijhuis, 1996).

These case studies show several differences of behaviour between the stakeholders inolved, which can be seen by focusing on 'snap-shots' of critical moments negatively influencing the progress of the construction processes. This follows the recommendations of, for example, Schein (1985) and Sanders (1995), both of whom more or less recommend focusing on critical incidents in the practitioner's situation if one wants to learn what the influence is of business cultures and their representing human behaviour on daily (construction) processes. For this book a few of several available examples have been selected, described within the existing case studies, merely on the basis of access to in-depth experiences. There are other examples of situations in which the progress of such ongoing construction processes is influenced negatively, but these case studies were selected because they exemplify daily complex situations between several stakeholders involved within construction processes.

Although originating from a broad scope of available case studies and experiences, the selected case studies in this book focus on critical moments during construction processes, in order to learn from them. In this respect, it is necessary to mention that all the described case studies resulted in more or less satisfactory positive solutions for all the parties involved. The problems arising during (parts of the) construction processes, and the solutions implemented, are a valuable source of experiences and lessons, acting as serious 'food for thought'.

The aim of including the case studies is to analyse them with the focus on how to avoid disturbance or delay of the overall construction process involved. The construction process is considered as the total process chain, consisting of the phases from *initiative* and *design*, through *tendering* and *preparation*, resulting in *construction*, until *delivery*, *use* and *re-use*.

All the described case studies are based on snap-shots of critical moments within their total construction process, positioned within one or more of these phases. In other words:

Focus in the case studies is mainly on how the analysed behaviours of the stakeholders involved within the specific phases of the construction process, and differences between them, were influencing the progress of the total construction process.

To give a condensed background information and overview of the projects involved, the authors have chosen to follow the approach used previously by them in analysing other projects (Tijhuis and Fellows, 2003), represented in separate paragraphs as follows:

- project description;
- stakeholders involved;

- differences in behaviour of the stakeholders involved;
- conclusions and recommendations: lessons learned;
  - *Contact* (culture);
  - *Contract* (project organization);
  - *Conflict* (technology).

The authors are grateful to the specific parties involved, for their openness in sharing their experiences and information about the specific project cases.

Case study 1: Developing a complex inner-city project

This case study represents a description of a project located in a regional town in the Netherlands. The project was characterized as an inner-city refurbishment site, with several additional complexities such as different initial site-owners, small building site, difficult project logistics and high time-pressure during construction. This case study is part of an earlier in-depth analysis, also carried out and published by the authors of this book (Tijhuis and Fellows, 2003), and gives an example of how complicated even smaller construction projects can be, especially because of the differences in behaviour of the stakeholders involved.

Project description

The project consists of two main types of building:

- two shops and/or offices on the street level;
- six apartment blocks, divided into three apartments on two storeys.

As it is located in an old city centre in a middle-sized Dutch town, it has some difficulties not only in logistics but also in legislation procedures because of the specific influences of the historical location. The location is represented schematically in Figure 4.3 (Straetement, 2002).

The building itself is a modern contemporary-style design, based on a concrete structure with outer walls of masonry in different colours and styles. It has three floors, with offices/stores and parking facilities on the ground floor and apartments with terraces on the first and second floors (Straetement, 2002).

Stakeholders involved

In the project, several stakeholders were involved. The following are mentioned, including their roles:

- **land-owner**: a private party, who sold the site to the developer;

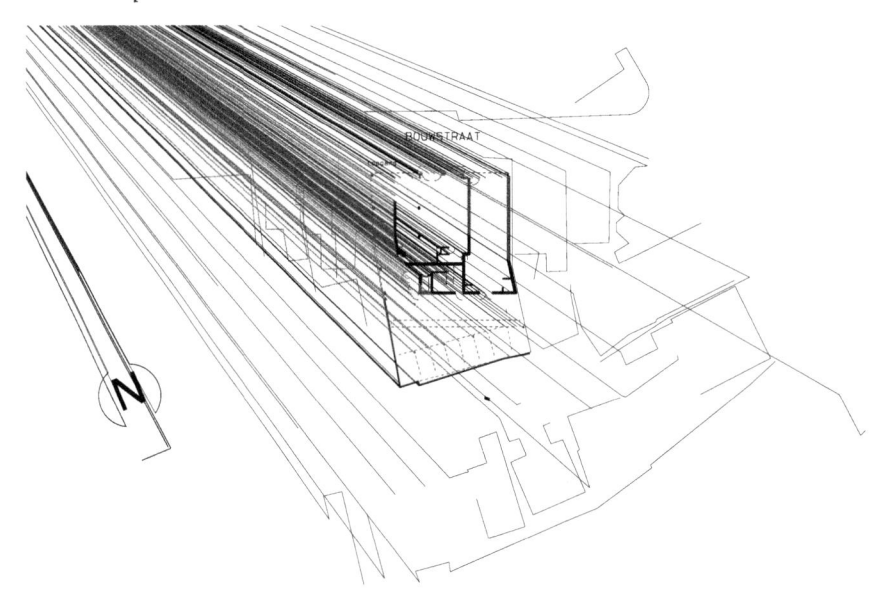

Figure 4.3 Schematic situation of the project location (Straetement, 2002).

- **developer**: a professional partnership between a developer, experienced in developing this kind of projects, and the architect, who knew the land-owner;
- **contractor**: the building company, without any influence in the development risks;
- **architect**: design and also partner at risk within the development combination;
- **consultants/engineers**: professional advisors in legal, financial, organizational and technical issues;
- **clients/investors**: non-professional and professional parties, buyers, owners and end-users of the building;
- **municipality/city council**: public party, proceeding with the legislation procedures;
- **neighbours**: private parties, worried about the influence on their living environments.

Differences in behaviour of the stakeholders involved

Although it quite often happens that the progress of projects is disturbed, one can never find full acceptance of this aspect in construction. As time is money, parties want to proceed as quick as possible. Time management plays an important role in such situations, but, especially in relation to legislation processes, one cannot use other means than just waiting and hoping the local governments proceed correctly. Unlike in project teams on sites, for example, one cannot use incentive programmes to speed up procedures (Laufer *et al.*, 1992), as this means of influencing public parties, is forbidden by, for example, the governmental 'Commissie Vos', as in a recent investigation of certain branches of the Dutch construction industry (Vos *et al.*, 2002). So, one can only wait for governmental or municipal action to proceed as described and organized in national, regional or local regulation frameworks.

Looking at the Dutch situation, this mainly introduces the large influence and degree of participation of parties in the immediate physical neighbourhoud of the site, giving them the official opportunity, based on the applicable procedures, for complaining within the legislation process of a building project. This really can take some time, or even stop and/or block the whole planning and/or realization process of a project.

Also in relationship to this, in the research of, for example, Hofstede (1980) the Dutch business environment in general is characterized as quite individual and with a low uncertainty reduction level. One can see such a basic attitude of Dutch business culture as focusing foremost on getting commitment and symbiosis between parties; in general it is also often considered as the 'polder model'; official legislation procedures also represent this attitude, by giving the individuals considerable rights and influence in them. Although on the one hand it is a positive structure, on the other hand it often nearly paralyses decision-making processes.

Within the legislation procedure of this project, the neighbours were strongly complaining about the plans. They had not expected that anyone would ever build a project opposite their apartment buildings on the other side of the street. That was just the problem: the project was immediately opposite their existing apartment building, which had been built about seven years earlier.

However, in those days the municipality also pointed out in its official town-planning documents that the extension of apartment buildings along this street would fit into their plans. So, there were no complaints from the public side, only from the private side, although there were also some neighbours who might be interested in working with the developer to extend the project in the near future. However, they also displayed typical behaviour, described below in more detail, and are numbered here as 'neighbour-groups' as represented schematically in Figure 4.4, mainly representing the non-professional parties involved in the process (Straetement, 2002).

In particular the following groups of neighbours and their behaviour were part of the overall construction process:

> *Group A*: This group were complaining about the unexpected building project opposite their apartments.

86 *The practice: international case studies*

Group B: Although they were satisfied when they sold the site for the project to the developer, this group were still a little difficult, as their remaining home just beside the project needed an extra parking place in the garden, with an entrance from the new building site; this was quite difficult to accommodate because there was little space left.

Group C: This group was completely satisfied, having gained, for example, a new garden wall in their backyard, and also having sold a small piece of land to the developer of the project.

Group D: This group was a rather 'difficult' party. Although they did not complain, they were still quite 'difficult' during construction of the project. For example, they were unclear about selling their site to the developer, making it difficult to discuss and/or negotiate with them. However, their site was still interesting for the developer as a possibility for extension of the planned project in the near future.

As should be clear, a good dialogue between the developer/team and the neighbours involved should still be worked on seriously, as Van Riemsdijk

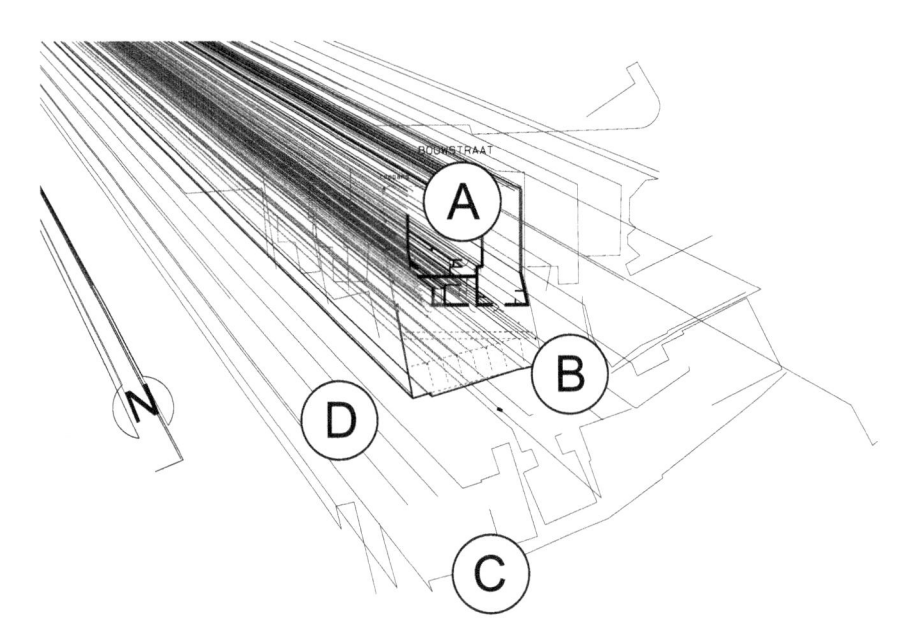

Figure 4.4 Groups of neighbours within this project, as parties involved in the process (Straetement, 2002).

also described in his research on large projects/companies and events with a big impact in society (Van Riemsdijk, 1994). This could at least reduce the neighbours' fear of such projects and the disturbance of their neighbourhood.

Besides these parties, as more or less 'non-professional', the professional parties involved in the project represented in general the following behaviour:

- **Land-owner**: As a private party, he was quite clear and strict in his approach to selling the site. As he knew the local architect, the architect claimed a sort of 'controlling position' when the developer bought the site.
- **Developer**: Given the position of the architect, the developer hardly could do anything other than form a joint venture with the architect in the buying of the site.
- **Contractor**: After a tender procedure, he was awarded a normal traditional building contract, based on the tender documents. In this case he acted as a reliable party.
- **Architect**: As he had a dual position in the project (co-developer and architectural designer), he continuously moved between these two key issues: (positively) improving profit by the developer and (negatively) increasing the payment for more design tasks by the architect. That resulted in a quite difficult way of proceeding.
- **Consultants/engineers**: Owing to the long delay in legislation (while the neighbours used their right of complaining/influencing the procedures; see above) they had to adapt their plans several times. However, they kept their behaviour in accordance with their professional role, and were willing to collaborate within the team.
- **Clients/investors**: As these parties were not involved during the legislation procedure and preparation phase of the project, they hardly influenced the process. However, there were, of course, different wishes for the definitive interior designs and so on, but that was not really a problem because of the process.
- **Municipality/city council**: As a public party, it proceeded with the legislation procedures. Although this sometimes led to difficulties in the process, it was quite willing to accept the plans. Only a small complaint in the matter of the architectural style of the first designs caused some problems. However, further on in the legislation procedure, the officially accepted complaints by the neighbours of the site really led to delay in the procedures.

It became clear that the behaviour of two main parties specifically was the main reason that the running process was disturbed.

88 *The practice: international case studies*

- **Architect**: mainly on account of his dual interests in development and design.

    Comment: As a member of the project team, his role should be to encourage rather than disturbing the running process. Therefore, his behaviour was felt by the other team members as really frustrating, and acted more or less as an 'internal block'.

- **Neighbours**: mainly because of their legal right to complain against legislation.

    Comment: As they were not members of the project team, their behaviour could not be expected to be specifically positive or negative, and they therefore acted more or less as an 'external block'.

Based on the 3C-Model™, one can represent the above situation as 'types of behaviour', being:

- positive (i.e. speeding up the running process);
- negative (i.e. slowing down the running process).

Moving forward in the process from the contact phase towards the conflict phase, it can be seen that the attitudes of the parties involved stayed more or less the same; in particular, the architect and the neighbours kept disturbing the overall construction process, instead of acting collaboratively to find a satisfactory solution to maintain progress. In Figure 4.5 this is represented schematically in the three phases of the 3C-Model™.

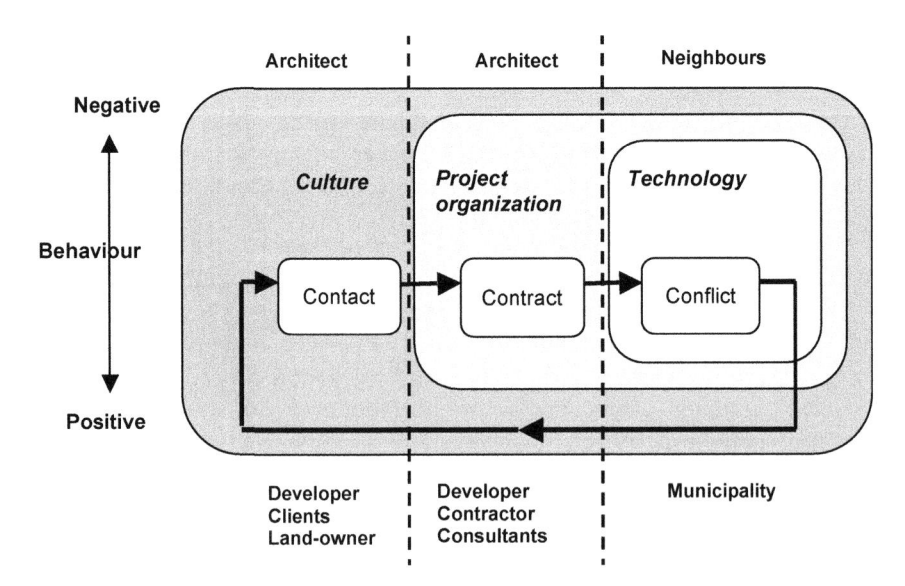

Figure 4.5 Types of behaviour (positive and negative) of the parties involved in the three phases of the 3C-Model™.

The practice: international case studies 89

So this quite small project was seriously delayed by specific differences in the behaviour of parties involved, which had not been recognized seriously when establishing the project team in its specific environment.

Conclusions and recommendations: lessons learned

The differences in behaviour could on one hand be seen as 'just differences', but on the other hand (mainly from a viewpoint of improving processes in construction) they implicitly represent some lessons, which are described below, based on the frameworks of Figures 4.1 and 4.5.

Contact (Culture)

LESSON 1

As the architect was already a member of the project team from the outset, he should behave according to the goals of the team, that is 'keeping the overall construction process going'. This led to lesson 1:

> A serious attention to the culture (behaviour, reputation, etc.) of possible parties involved when establishing the project team should therefore be recognized within this selection process. The 'match' with its project environment should be taken seriously. Trying to understand team members' 'hidden agendas' as well as their 'official agendas' is of great importance.

LESSON 2

The neighbours came into the process only when legislation started, that is after the team members had signed contracts and so on. Therefore, they were, on one hand, not 'part of the deal', but, on the other hand, were still in a position to disturb the whole project, owing to their official right to influence the process. This led to lesson 2:

> Good communication at the front end of the process, involving all the stakeholders, would possibly have improved the acceptance of the project; at least, if the architect and developer had accepted their influence at an early stage.

Contract (project organization)

LESSON 3

As the generally accepted and assumed role of an architect is a neutral one, as a design specialist, he should not assume additional roles in one and the same project.

Despite his generally assumed 'neutral' role, the architect had obviously two separate main roles within this project: designing and co-developing. This led to a very difficult situation in that he had more than one goal: working on an attractive design with high fees for himself and realizing a high-profit project. Lesson 3 is thus defined as:

> Keeping a clear separation of the several roles within projects, divided among the parties involved, is always possible. However, when deciding this in the contract, each of the separate parties should have roles, being related to at most one side of the 'three sides of the table' (i.e. client/developer-related, or neutral-based or contractor/subcontractor-related).

LESSON 4

A joint venture between the developer and the architect led to differences in behaviour not only because of their differenct 'agendas', but also because of differences between their field of experiences and 'branch-culture'. Lesson 4 is therefore:

> Choosing and deciding about team members for a (building) contract should be not only on a legal or financial basis, but also based on the experiences and 'culture' of the parties involved.

Conflict (technology)

LESSON 5

The neighbours were complaining about the plans presented to the municipality during the legislation procedure. Although in the end their complaints were not upheld, one cannot be sure that their behaviour during the use and 'life cycle' of the realized project will be amenable. Thus, communication with them should still be organized in an effective manner by informing them about the technical aspects, levels of expected noise and nuisance during construction realization and so forth. This should take away or at least reduce their fear about the 'new' situation, especially during the construction process. Lesson 5 is defined as follows:

> Besides the project team members, an open and structured communication (a 'dialogue') during the total construction process is an effective instrument for taking away certain fears of stakeholders (with active and less active roles) in projects. This certainly can reduce the risk of disturbing projects in an early stage, although it could require extra attention and energy on the part of the team members. But that's worth it.

Resumé

Working in construction processes generally implies that the parties involved should have the same goals – initiating, designing or realizing – thus delivering a project within the basic conditions negotiated between the public and private parties. However, as a major result of the experiences described in this case study, one should be aware that there is a need for growing awareness of the fact that, given differences in roles, for example as a result of the basic project contract, the behaviour of the different parties can seriously lead towards conflict situations. Thus, taking more care in establishing project teams and 'matching' them with their environment, looking in more detail at the basic behaviour (culture, reputation, etc.) of the stakeholders involved, can reduce the risk of unsatisfied contracts and even avoid conflicts. In particular, such an approach may lead to a substantial improvement in construction processes.

Case study 2: Construction of rationalized terraced housing

This case study describes a project of terraced family housing, located in a large city in the Ruhrgebiet region of Germany. This region is undergoing a great deal of development in terms of housing upgrading, expansion and modernization. The project is part of a new inner-city development (a so-called *Wohnsiedlung*) of new houses built on the basis of a standardized concept. The clients were private German individuals or families with a direct purchasing agreement with the German developer.

Project description

The site is located along a new road that is an extension of a German urban area. The basis was a development plan for several blocks of houses (i.e. houses in a row). These houses were built as part of a total package deal by a German contractor, which was also responsible for the detailed design (a so-called *General-Übernehmer*). The regional developer (a *Bauträger*) also handled the selling and liaison with the clients. Figure 4.6 shows part of the housing project during the construction phase.

Part of the problem in this project, and the main reason for disturbance in the construction process, was the discovery during construction of a leak in the cellar of a few of the houses. If such problems only become apparent after completion, for example because of heavy rainfall and/or rising groundwater levels, then repair is often difficult and expensive. That was the case in this project: the final finishing works (gardens etc.) were completed, the houses had been handed over to the end-users, and then it was discovered that some of the cellars were leaking. Of course, this resulted in dissatisfied end-users and an unhappy developer. It was obvious that the technical specifications, work preparation and realization were the responsibility of the contractor; especially because the type of contract which was used here was the so-called

92 *The practice: international case studies*

Figure 4.6 Example of a part of the housing project during its construction phase.

General-Übernehmer-Vertrag (Beck and Herig, 1997). Within this contract, the contractor is responsible and liable for the total design, technical details and functionalities, and engineering, as well as the total realization of the project. In such situations, in general, the developer (as the direct client of the contractor) specifies only the total concept and basic design, thus bearing the commercial (market) risk, whereas the contractor bears the total technical (detail) risk.

Stakeholders involved

In the project, several stakeholders were involved. The following are mentioned, including their roles:

- **developer**: a professional German company, experienced in this kind of project, and accustomed to working with several architects to develop (standardized) housing concepts;
- **contractor**: a German building company, responsible for the technical details and preparation of the construction (working drawings, realization, etc.);
- **architect**: a German architect delivering the conceptual design, and responsible for the coordination of the planning permission;
- **consultants/engineers**: professional German advisors in legal, financial, organizational and technical issues;

- **clients**: non-professional parties, mainly families, as buyers, owners and end-users;
- **municipality**: public party, proceeding with the legislation procedures, and seller of the site to the developer.

Differences in behaviour of the stakeholders involved

In this project, the leaking cellars led to a dispute in which the contractor was unwilling to accept responsibility and liability. However, at the beginning of the project and during the tendering stage, all parties involved were intent on achieving the best results. This was reflected, for example, in the choice of contractor. The party eventually selected was successful because of its ability to organize and coordinate the total building process, as proven by its track record in the realization of similar housing projects in the same region. Formally it was also certified according to ISO-9001/9002, which functioned as an extra proof of the contractor's experience and knowledge in its field of activities.

However, when after completion some of the cellars appeared to be leaking, and the developer started to blame the contractor by pointing to its responsibility and liability for this technical failure, the contractor started to defend itself, not only by naming and blaming the subcontractor (who had in the meantime gone bankrupt on another project) but also by pointing out the possibly unreliable soil conditions and groundwater levels. Figure 4.7 shows the construction phase when additional waterproofing work on one of the cellars was being carried out.

All the discussions led to a prolonged dispute during which claims against the contractor were made by the developer, and claims against the developer were made by the end-users (Felstiner, Abel and Sarat, 1980).

This can be recognized as characteristic of the general German business culture, as represented by Hofstede in his international research. He characterized German business culture as having a medium power distance and a high level of uncertainty avoidance, experienced as a 'professional bureaucracy' (Hofstede, 1991: 191). In fact, these characteristics may be some of the reasons that conflicts within German construction industry often lead very quickly to large legal or juridical diputes and court cases; there appears to be very rare for there to be any kind of informal negotiations or mediation between parties before going to court, as also described previously by Tijhuis (1996).

In fact, the end-users as 'non-professional' and the others as the 'professional' parties involved in the project represented in general the following behaviour:

- **Developer**: Claiming that the contractor should solve the problem, especially by pointing at its responsibilities and liabilities, because of the contract type (*General-Übernehmer Vertrag*); however, the developer

94 *The practice: international case studies*

Figure 4.7 Work phase in which additional waterproofing work is being done on one of the cellars.

was, in relation to the end-users as its clients, responsible for the overall result as initiator of this project.

- **Contractor**: Despite its existing quality systems, something had gone wrong during the construction process, perhaps because it used several subcontractors during the cellar construction, which was critical. The contractor's defence focused on the (theoretical?) possibility that the soil conditions and groundwater levels had changed 'at some point', thus changing the specifications needed for the cellar construction.
- **Architect**: As the architect was responsible only for the conceptual design and the coordination of the building permission, he was not a part of the dispute.
- **Consultants/engineers**: In this type of contract, coordination of these professionals is the responsibility of the contractor. The consultants/engineers were required only to check specific aspects of the work. Thus, they they were involved this dispute only on a consulting level. However, the role of the construction lawyers and technical experts and arbitrators (*Gutacher*) was quite intensive.
- **Clients**: As the end-users, the clients claiming against the developer, from whom they had bought their houses.
- **Municipality**: As the public party, the municipality was not really involved in this dispute, although in the discussions about groundwater levels and so on there was some influence from infrastructure realization in the neighbouring areas.

It became clear that the behaviours of specifically three main parties were the main reasons for problems in the construction process.

- **Developer**: Because of its choice of the specific type of contract and a certified contractor, the developer trusted that the construction process would go smoothly.

    Comment: It seemed that the developer was after all relying somewhat too much on trusting the contractor. However, at the beginning of the construction process the relationship between developer and contractor seemed to be based on a good mutual understanding, well-known experiences in comparable previous projects and so on. Because the developer wanted to keep the relationship with its (complaining) clients as end-users as good as possible, it put them in a difficult position: in fact 'stuck in the middle' between the contractor and the end-user.

- **Contractor**: Although having sufficient experience and a well-implemented quality system, it pointed out that working with different parties and subcontractors increased the risk of construction failures. From the moment of the dispute, every attempt was made to avoid any liability for them, with no focus on the complaining end-user at all.

    Comment: It was shown here that it is extremely important to control and review contracts, daily work reports and so on, especially when working with different parties. But even a control system is no guarantee of a perfect result. One should also communicate not only with the direct contract partners, but also with the stakeholders, to keep the trust involved. It often seems to be a matter of finding a balance between trust and control (Tijhuis, 2004).

- **Clients**: As the clients were also the end-users, they were immediately confronted with the cellar leakage. However, they also immediately relied on construction lawyers, instead of first trying to seek a non-juridical way of solving the problem.

    Comment: It was understandable that the clients, as non-professional parties, used the juridical way of solving the problem. As Felstiner, Abel and Sarat (1980) also explained, this is an oft-used way of handling disputes, and also influenced by the (social) environment. Especially nowadays one can still see a strong increase of such practices, although an early statement about the role of lawyers, written in Shakespeare's play *Henry VI*, still gives people today food for thought: 'The first thing we do, let's kill all the lawyers' (Shakespeare, 1964: 111). This is quite interesting, if 'translated' more to the (often disturbing?) function of legal practices nowadays, especially in the construction industry.

Based on the 3C-Model™, one can represent the above situation as 'types of behaviour', being:

- positive (i.e. speeding up the running process);
- negative (i.e. slowing down the running process).

96 The practice: international case studies

Moving forward in the process from the contact phase towards the conflict phase, there could be seen changing attitudes of the parties involved; in particular, the relationship between the contractor and the clients changed from a positive collaborative one into a negative disturbing one. Finding a satisfactory solution to keep the progress became indeed quite difficult. In Figure 4.8 this is represented schematically in the three phases of the 3C-Model™.

After all, this project did result in an unsatisfactory situation with respect to a few of the delivered houses. However, the developer and contractor agreed in the end to share the costs of repairing the cellar leakage. Nevertheless, the case study pointed out that trusting parties is a good thing, but checking and reviewing them is often still necessary.

Conclusions and recommendations: lessons learned

Contact (culture)

LESSON 1

As the developer and contractor knew each other quite well from the beginning, and there was sufficient availability of quality systems, there was an increased trust in the technical capacity of the contractor. However,

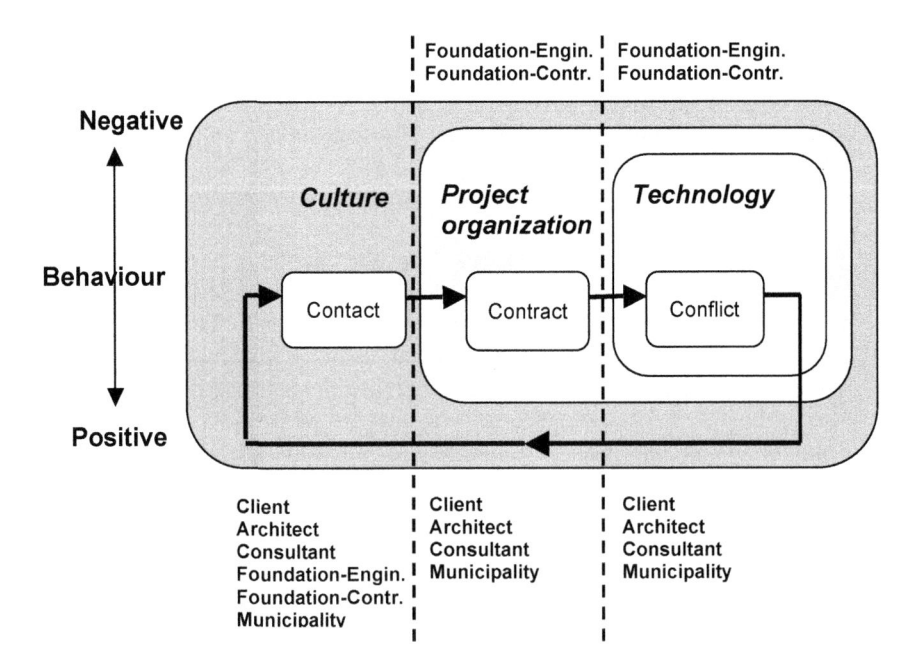

Figure 4.8 Types of behaviour (positive and negative) of the parties involved in the three phases of the 3C-Model™.

controlling the process was obviously therefore made somewhat 'incomplete'. Thus, lesson 1 can be defined as:

> Trust is good, but control is better; even when parties do know each other quite well, there is still a serious risk of changing behaviour due to the influence of a changing situation within the construction process. Nevertheless, too much control can be counter-productive because of the risk of frustrating the parties involved. Therefore, there is especially a need of a good balance between trust and control.

Contract (project organization)

LESSON 2

Choosing the right type of contract is essential. This means that there should be a cultural fit between the parties involved, not only regarding their behaviour, but also and especially regarding their way of handling project organizations and their incorporated contracts, in particular when the contracts incorporate a complex responsibility and liability structure. Therefore, lesson 2 is defined as follows:

> The decision for choosing the type of contract should be based not only on the characteristics of the project itself, but also and especially on the balancing of the involved participants' characteristics. Parallel to this, the decision on accepting the contract should be made not only on the basis of the project characteristics, but also and especially on the basis of the 'suitability' for one's own organization.

Conflict (technology)

LESSON 3

Technical problems during projects are often the basis for conflicts during the construction process ('the proof of the pudding is in the eating'). However, the way to handle these conflicts differs quite a lot, and varies in general between the personal and the juridical ways of solving them, with all the complexities involved in these ways. However, to be able to handle the conflicts as well as possible, it seems that the emerging trend in the professional as well as private world internationally is towards the juridical method, even in countries and cultures that were originally used to the personal way. Therefore, lesson 3 is defined as follows:

> Although one is more comfortable with a personal way of handling conflicts, one should be prepared to accept that the juridical way will often become the final means; this when working with professional as well

as private parties. This may also be an extra reason to be very selective during acceptance or rejection of projects and their accompanying contracts and business partners.

Resumé

Although often professional and private parties are participating in construction processes, there can still be differences in 'typical' behaviour between professional and private parties. However, regarding international trends, one can see an increase from personal towards juridical handling of conflicts. Apart from the project characteristics, this makes it even more difficult and necessary to know with whom one is dealing, and under what conditions, from the very beginning; investigation of track records, experience and know-how is a large part of this. This leads to a growing need for a people-centred approach, instead of just a technology- or process-driven approach. In fact that is not a strange thing, because *construction has been, since the earliest times, and is still a business of people.*

Case study 3: Subcontracting infrastructural and foundation works, a Polish case study

In this case study the preparation and realization of an industrial facility in the central region of Poland is described. During the late 1990s, after the fall of the Iron Curtain, Poland and other Central European countries experienced a strong increase in development and construction projects, following the modernization of their economies. During that period, the market rapidly opened up, and these countries were also working to align their regulations with those of the European Union (EU), in preparation for joining the EU (Beck, 1998a,b). This case study originates from that period. The project described is part of a series of nearly identical facilities for one international Western European client, in the phase of establishing its (international) business in the Polish market.

Project description

The described site is located on the edge of a new large industrial development zone within one of the larger cities in central Poland. As this project was one of the first developments there during the period between *c.* 1995 and *c.* 2000, the municipality had appointed this one as a kind of example for further developments to come. So the public parties were very keen to achieve the best result possible.

In general the project consisted of a large garage complex in the first development phase, including offices and facilities. In the second development phase housing for some of the key employees was added.

Figure 4.9 shows an aerial view of parts of the building foundations and parts of a cellar during construction.

Figure 4.9 Parts of the building foundations and parts of a cellar during construction.

The general concept of the building was a large concrete basement and cellars, on top of which a prefabricated steel building was erected. As the groundwater level of the site was quite high, the opinion of the local specialist engineers, involved in the legislation process, was pertinent: according to them, the site was not suitable for a standard concrete cellar. Thus, they advised that the natural ground level of the site be raised quite significantly, by more than two metres; so instead of just making the concrete cellar structure more watertight, for example by using thicker concrete walls or high-tensile prefabricated concrete structures (which was and still is quite common in other parts of Europe, and is also used in, for example, comparable German construction standards; DIN, 1988), they recommended (indeed, insisted on!) raising the site level by supplying extra sand, which meant a lot of extra cost. This meant an interesting start for a difficult negotiation between the Dutch architect and its process manager, the client and the local foundation engineer and foundation contractor. This is outlined during the following analysis.

Stakeholders involved

In the project, several stakeholders were involved. The following stakeholders are mentioned, including their roles:

- **foundation contractor**: the local building company, focusing on realization, not responsible for the risk on the design and specifications of the technical details;

- **foundation engineer**: professional local party, responsible for the risk on the design and specifications of the technical details, and especially active during the legislation procedures;
- **architect**: responsible for conceptual design and the coordination of the building permission including the process management;
- **consultant**: professional advisor on legal, financial, organizational and technical issues;
- **client**: a professional company, experienced in other projects of this kind for its own use;
- **municipality**: public party, proceeding with the legislation procedures and seller of the site to the client.

Differences in behaviour of the stakeholders involved

This project was experienced by the international parties involved as an example of how local networks are being used to try to control international clients, thus showing several of the 'typical' risks of international activities in construction and development.

As the project was the third in a series of in total five quite identical facilities, the client chose to use this project as pilot exercise for trying to get down the construction cost by optimizing the design concept and accompanying construction process as experienced so far during the realization of the previous two facilities (i.e. 'learning by doing'). At the same time, the client was trying to replace the existing contracting parties involved so far (i.e. members of a pool of business partners, often working for the client in other such projects elsewhere), with new, solely local, parties. That was an interesting opportunity, but also increased the risk of communication and coordination errors and the like (especially because of changing the contracting parties involved).

During the design phase and the subsequent preparation phase, the negotiations with the (partly new) stakeholders involved went on quite slowly. This appears to have been because the available standardized basic design of the cellar construction, in relationship to the soil conditions, caused the local foundation engineer and foundation contractor to react as follows.

According to them, the available design was not suitable for the soil conditions, especially because the groundwater level was quite high on this site (i.e. about 0.5 m below the surface of the earth). However, the reason was not made quite clear by them (reports were relatively infrequent), which was quite unsatisfactory for the client, architect and consultant, especially because the type of 'standardized' and proven design chosen initially for this cellar was not unusual; moreover, there were enough examples of projects where this design had been used before, some elsewhere in Europe, so why not use it in Poland? Such construction could easily be made water-resistant, if built properly.

Nevertheless, the Polish parties involved insisted on implementing their suggested solution as follows.

Instead of digging the hole for the large cellar and foundations (i.e. digging in total about 2.5 m deep, over an area of about 400 m², which equates to about 1,000 m³ earth movement; this would result in the cellar floor being located about 2 m below groundwater level), they suggested raising the total site with a supply of additional sand. According to their proposed solution, it would be possible then to build the cellar floor above groundwater level. However, that meant that the whole site (*c.* 15,000 m²) had to be raised about 2 m, resulting in a requirement for *c.* 30,000 m³ of additional sand net (i.e. *c.* 35,000 m³ gross)! So it meant a real increase in construction costs not only for earth removal, but also for the purchase of additional sand itself, a rather strange situation, and unacceptable to the client, architect and consultant.

Figure 4.10 shows the whole building site, which is in a quite 'rural' location.

The situation gradually developed further into a quite strange situation, described briefly below.

When the parties were negotiating about their proposed solution, the Polish foundation engineer pointed to the fact that the foreseen extra sand supply should be of special quality, only to be found within a large sand-pit about 30 km away from the site; and this, although there was quite a big sand-pit in the neighbourhood, circa 5 km away, which was also used for the supply of additional sand for other construction works in the neighbourhood.

So this attitude of the foundation engineer gave a somewhat strange feeling to the client, architect and consultant. Why not accept a cellar built below groundwater level? And why be so sure that the sand from a specific

Figure 4.10 The whole building site in its 'rural' location.

102 *The practice: international case studies*

sand-pit is the only suitable quality for this site? Are there maybe other 'typical/bureaucratic' reasons for this behaviour?

Looking more closely at certain behavioural characteristics as representatives of business cultures, the international comparison studies from Hofstede are often used. However, Hofstede did not document Polish business characteristics (Hofstede, 1980); furthermore, during the research period (c.1980), Poland was still a member of the communist bloc.

Nevertheless, after the collapse of the USSR, Poland followed a strong economic growth trajectory. If one deduced that Poland, as a neighbour of Germany, would also demonstrated the German documented 'professional bureaucracy' characteristics, as described by Hofstede, this might give a partly correct representation of the Poles' business behaviour.

The correspondence is only partial because, as a high-growth country undergoing (economic) transition, Poland also seems to exhibit a kind of opportunistic approach – what others might call 'cowboy behaviour'.

In practice, the client experienced the Polish business behaviour as combining quite a low level of uncertainty reduction on one hand (for example by Polish people taking up several new challenges as entrepreneurs in their newly established home market), but on the other hand also showing quite a high level of uncertainty reduction too (for example by Polish people being quite inflexible about changing their initial viewpoints and/or habits). Altogether it is thus a quite challenging business environment. This might also be fuelled by the fact that the period of growth of the Polish ecnomy was combined with and indeed influenced by a search for new ways of doing business.

Nevertheless, the exposed behaviour of the foundation engineer led to the situation that the client, architect and consultant became now even more alert, and were therefore acting as follows:

1. They asked for a second opinion on the cellar construction design, asking another internationally active structural engineering office to recalculate the initially proposed structural design.
2. Parallel to that, they took samples from the sand of the site and from both sand-pits, and took them to a laboratory in the Netherlands. There the samples were tested and compared regarding their characteristics and suitability for the requested situation.

After both actions, the conclusions were as follows:

1. It proved to be the case that, according to the usual international standards in use for cellars built below groundwater level, the water pressure on the cellar walls did not create a need for any additional steel bars or for an improvement of the suggested concrete or cement quality whatsoever, but would only require good quality control during the realization.

Eventually, it was recommended to add some specific waterproof coatings on the outside of the cellar walls.
2   The other conclusion was also interesting, because the characteristics of the sand on the site did not differ much from those of the sand in the two sand-pits. In fact, just adding and mixing with some stabilizing material such as cement could considerably increase the stabilizing quality of the surface layers of the site.

Both conclusions meant that no special sand supply was needed, because the cellars could still be built 'under groundwater level'. If it had still been decided to use sand, then even the sand in the neighbourhood was suitable for this.

These results of the contra-expertise strengthened the feeling of the client, architect and consultant that the Polish parties involved were acting as if 'playing a game with them', showing a very strict and bureaucratic behaviour, trying obviously to make their profit as big as possible. However, additional information came after 'digging' somewhat further into the files and registrations of both Polish parties. The architect and consultant found out that the Polish foundation contractor was of course a very respected party in the region (as was also the foundation engineer) but he was also co-owner of the specific sand-pit, located about 30 km from the site! This encouraged the client to believe that solely commercial interests were the reason for the contractor's behaviour. However, although these commercial interests seemed to be the real reason for the 'typical' bureaucratic behaviour of the Polish parties, it still did not give a feasible solution for the client. The project was gradually coming under time pressure (which is a quite common reason, in construction projects, to 'force' parties involved into a certain solution or direction). So, keeping into mind that especially the client's interests had to take priority, it resulted into a quite interesting move during the negotiations, as follows.

As the client was leading, it meant that on the one hand he wanted value for money, but on the other hand he did not like to pay to conform with 'overperformanced' guidelines or requests from the Polish foundation engineer. It somehow seemed that both these newly involved Polish parties were acting as a typical small team in a kind of existing group dynamic, not really leading to a high performance for their (external) client (Katzenbach and Smith, 1993). Nevertheless, as this client's core business was the dealership and maintenance of heavy-load trucks and trailers, this fact pointed to a quick and feasible solution, as follows:

- For the mid to long term, the foundation contractor needed to upgrade or replace much of his equipment (trucks, concrete mixers, etc.).
- As it was a growing market (i.e. the period *c.* 1995–2000), such special equipment was quite expensive.

104 *The practice: international case studies*

The above issues resulted in the end as follows.

The client took the lead, and made a deal with the foundation contractor to deliver him new trucks and concrete mixers for a quite reduced price. Parallel to that, he was allowed to make the foundations and the cellar according to the internationally normal standardized way (which was also accepted by the Polish municipality regarding the requested building permission). Most importantly of all, they agreed to have the cellar built below groundwater level, which meant that the additional sand supply was not necessary at all.

Looking at the total case, in fact the professional parties involved in the project represented in general the following behaviours.

- **Foundation contractor**: Was in fact for long time acting 'behind the scenes', but it became clear that he had a strong commercial interest in the 'typical/bureaucratic' behaviour of the foundation engineer.
- **Foundation engineer**: Although a recognized professional party, he proved not to be acting independently. The informal relationship he had with the foundation contractor meant a long period of lack of clarity and disturbance of the project's progress.
- **Architect**: Was well known to the client, but was also depending on the client's decisions for go/no go into this 'typical' Polish market during this period.
- **Consultant**: As a professional advisor he was trying to keep an overview, but together with the architect was also misled for a long time by the foundation engineer and foundation contractor.
- **Client**: A professional company, with the power to deal according to its own interest, quite good in negotiating on the basis of mutual interests and creating 'momentum' for this.
- **Municipality**: As a public party quite bureaucratic, and merely following the results of the foundation engineer's work and investigations, especially because the Polish parties involved still had a quite good reputation.

It became clear that the behaviours specifically of three main parties were the main reasons for impeding the construction process.

- **Foundation engineer**: His inflexible and bureaucratic behaviour during the structural design phase resulted in a large delay during the preparation for building permission.

    Comment: Although the client and his architect and consultant had worked earlier in the Polish market with comparable 'standardized' building concepts, the decision to choose a 'new' foundation engineer and foundation contractor proved to result in an unsatisfactory situation: there was obviously a 'mismatch' between them and the pool of existing business partners working for the client already. Parallel to this, the typical collaboration between the foundation engineer and foundation contractor, as both 'new' Polish parties in this pool, resulted in an uncontrollable situation for the client, in which morale was unpredictable (Kahn and Katz, 1953).

- **Foundation contractor**: The way he worked together with the foundation engineer 'behind the scenes' caused an unclear situation for the client, being also the reason for disturbing the construction process.

  Comment: The 'collusion' between foundation engineer and foundation contractor resulted in an unsatisfactory situation for the client especially. However, it should have rung some bells at the client's office that having two partners of the same nationality (Polish) starting more or less at the same moment as new parties in an existing pool of business partners will cause quite a strong risk that they will collude together, trying to reduce their initial risk and/or increasing their possible profit.

- **Clients**: The changing of the business partners caused an increased need to adapt the way to handle these business partners. However, the client was not alert enough at the beginning of this process, obviously only focusing on reducing costs, which led to a quite long negotiation process.

  Comment: Only focusing on reducing costs can be a good attitude but, as several practitioners in the construction industry know, it also has to do with the delivered quality. However, in this case it was also clear that having in-depth knowledge of one's business partner's capabilities is also necessary, especially in relationship to the circumstances and project characteristics.

Based on the 3C-Model™, one can represent the above situation as 'types of behaviour', being:

- positive (i.e. speeding up the running process);
- negative (i.e. slowing down the running process).

Moving forwards in the process from the contact phase towards the conflict phase, there could be seen changing attitudes of the parties involved; here especially the foundation engineer and the foundation contractor changed from a positive collaborative manner during the first contact phase into a negative disturbing manner during the discussion about the contracts and their proposed technology solutions. Finding a satisfactory solution to maintain progress became quite difficult. In Figure 4.11 this is represented schematically in the three phases of the 3C-Model™.

Despite everything, this project resulted in a satisfactory conclusion for the client, although the client was left with a bad taste about the behaviour exhibited by other parties during the negotiations. Focusing more intensively

106 *The practice: international case studies*

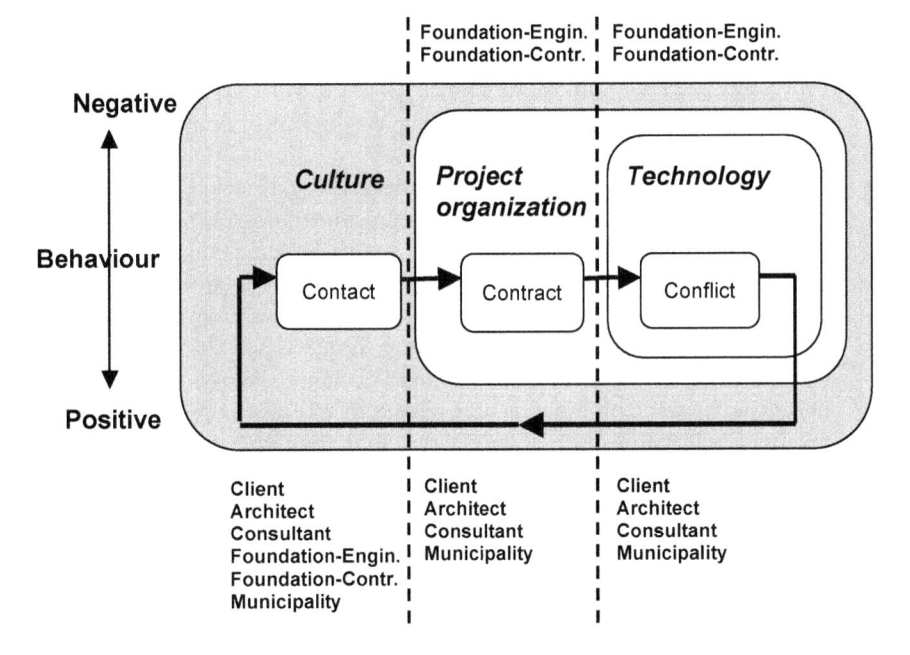

Figure 4.11 Types of behaviour (positive and negative) of the parties involved in the three phases of the 3C-Model™.

on reliability and keeping the match between the parties involved was from that point one of the leading issues for the next projects to come within their project investments.

Conclusions and recommendations: lessons learned

Contact (culture)

LESSON 1

Replacing the parties involved with new parties (as a tool for a team to promote competition between the existing parties involved), thus focusing on keeping a good price and quality, can prove to be a quite risky strategy within new and/or international markets. This is because a new party's behaviour may result in a negative influence on the whole team's behaviour. Therefore, lesson 1 is defined as follows:

> Competition within a team is good, to keep every team member's approach focused on the client's goals. However, before selecting and/or exchanging team members, one should carefully investigate the match between these party's attitudes, thus keeping a grip on the team (group) dynamics.

Contract (project organization)

LESSON 2

Although contracts play a significant role in (construction) projects, they do not guarantee the optimum performance of the parties involved. Before signing such contracts, it is recommended that one should try to determine the 'hidden agenda' of the parties involved. However, this is, of course, a contradiction in terms: how can one know another party's hidden agenda? Although in this case study the hidden agenda was revealed through learning-by-doing, it is obviously a continuous step-by-step process of being sensitive to weak signals, for example. Therefore, lesson 2 is defined as follows:

> Contracts are nowadays a necessary tool for getting projects realized. However, one should be very sensitive to the weak signals of the parties involved, during the process of negotiations before, during and after the signing of the contract. This especially because construction is still a people business, which means that project results cannot just be guaranteed by contracts.

Conflict (technology)

LESSON 3

This case study has shown that adhering to the same solution to a problem often results in a status quo: parties stick to their originally suggested or wished solutions, without considering totally different alternatives. In this case the client considered other interests – his own trucks business – which unexpectedly coincided with the foundation contractor's latent need for new trucks and concrete mixers. Therefore, solving that issue as part of the total solution for getting the construction process going again was the key to this result. Therefore, lesson 3 is defined as follows:

> During negotiatons for getting problems solved, one should especially keep the focus on those issues which connect parties, instead of focusing only on those issues which are just separating parties. This will increase the chance of finding solutions by routes and/or issues of mutual interest between the parties involved.

Resumé

Working in new and/or international regions is often a challenge for the parties involved. It is logical that one tries to incorporate as much as possible the local cultures within the team, but one should be aware that there is always a need to create a kind of match between those parties involved: not only a match between cultures, but also a match between the parties' interests.

Especially, creating insight into, and keeping a grip on, the hidden interests of those parties involved is a challenge for creating a real team, going for the best (project) result.

Case study 4: Tendering for developing a production factory

This case study describes the development of a production factory in the Istanbul region of Turkey. As Turkey is under consideration for several international industrial companies to expand and/or to relocate their production facilities to this country since about 2000, it has experienced a strong growth of investments. Especially the region around Istanbul attracts several of such international clients, because of its excellent location along the coast, creating logistically a good opportunity for establishing an export/import hub; and of course not forgetting the fact that the Turkish population is quite a young one, promising a strong growth of national consumption in the long term. However, although Turkey is quite near the Western and Central European region, it is still considered as being part of Asia and/or the Middle East, which is an intersting function as a kind of connecting 'gateway between the East and the West'. In another way it is still an interesting country too, because it represents in fact a bridge function between the Western European Christian cultures and the Asian and Middle Eastern Islamic cultures. The aforementioned region of Istanbul represents very literally this bridging function, connected by the famous Bosphorus strait.

Project description

The development project under consideration was expected to be sited in one of the industrialized zones, directly located outside the densely populated urban district, with good connections to roads, waterways and railways. At the time in question, the final site had not yet been definitively selected, but this issue was one of the discussions during the total development and construction process. In general the project can be described as follows.

The international client, with its headquarters located in North America and with subsidiaries in several parts of the world, specializes in producing industrial specialist supply-parts for several other industries. Because of its growing exports to Turkey, it wanted to establish its own new production factory there. For this, it invited development and investment consortia after a pre-selection phase, being the tender participants, to provide the client with proposals for turnkey solutions. The total requested offer was in general as follows:

- a total package of a gross rent per m^2 per year;
- for a rental period of at least 15 years;
- including a suitable site location (total $c.$ 15,000 m^2);
- plus a production factory (total $c.$ 10,000 m^2);
- including offices (total $c.$ 2,000 m^2).

For this, the client provided only a basic programme of requirements, including a basic plant layout, which had been designed by its British architect. This architect also conducted more or less the selection of the offers and parties involved, together with the client's operations manager responsible for Turkey.

Figure 4.12 shows the proposed building site, used as a basis for preparing the offer and site selection by the consortium involved.

As this building was in general not particularly complex regarding, for example, the requested prefabricated concrete structure, the complexities were more or less in the fact that the local production circumstances necessitated a fully temperature-controlled facility. Although not strictly a clean-room, it still needed certain air filters, air cooling and so on. For such large volumes this was quite a challenge for contractors and engineers. So the request for alternatives on the basis of the foreseen guidelines was still an interesting challenge for the selected tender participants. Although in Turkey's construction industry there are differences between parties involved (e.g. architects and engineers) regarding their professionalism and related way of behaviour (Akiner and Tijhuis, 2007), this was no reason for the client to choose, for example, for a British architect and Turkish and Belgian/Dutch consortium: the client just needed a feasible and competitive solution. How that process went is described in the following analysis.

Figure 4.12 The possible building site.

110 The practice: international case studies

Stakeholders involved

In the project, several stakeholders were involved. The following stakeholders are mentioned, including their roles, representing one of the consortia involved:

- **developer**: international Belgian/Dutch party, with experience in several similar industrial market segments; focusing on the total package deal including the acquisition of a suitable site, and leader of the consortium;
- **contractor**: Turkish party, with construction experience in such industrial facilities, including engineering; as partner within the consortium, responsible and liable for the technical risks after delivery;
- **investor**: professional Turkish international investor, active in several branches throughout the Middle East and elsewhere; actively participating in the concept and location selection, with a long-term vision, and also part of the consortium;
- **architect**: professional British party, having worked together with the client as a 'semi-internal' architect for many years; responsible for a smooth process to get the facility realized and running;
- **client**: a professional North American company, experienced in several such projects elsewhere, and represented by its local Turkish operational manager with support from its central international business department.

Differences in behaviour of the stakeholders involved

In this project it became clear that the mixture of different attitudes, also influenced by the several national and business cultures of the parties involved, led to a quite uncontrollable situation for all, especially because during the tender phase the client's market circumstances changed quite significantly, influencing all the other parties involved in one way or another, too. This meant a changing strategy for the client. In general the situation for this consortium and the client and his architect during the analysed tender phase was as follows.

The client had looked over its portfolio of existing business partners to see if there were suitable candidates to fulfil its existing need for a new production facility in Turkey. As its financial strategy had decided to go for a leasing agreement with a (developing) investor, the selection process was based solely on the following criteria:

- financial conditions (i.e. annual rental costs) in relation to the quality of the new facility;
- suitability of the location (i.e. geographical and technical);
- reliability/reputation of the business partners;
- speed of the total delivery time.

As the client already had contacts with Turkish parties because of its ongoing business operations there, Turkish parties were invited to make a proposal. Parallel to this, international parties were also invited to apply for this tender. Based on the availability of the information, this analysis is based on the experiences of these international parties within this process.

Another important characteristic of the situation was the following:

- The tender procedure itself was in fact an unofficial tender; it was mainly an invitation by the client to its well-known business partners to prepare a serious proposal for its planned new production facility.

This resulted in the situation that the procedure was in fact no procedure or agreed contract model at all; instead of this, it turned out to be a quite 'flexible' process, not based at all on well-known procedural contract models such as FIDIC (Fédération Internationale des Ingénieurs-Conseils), JCT (Joint Contracts Tribunal) and NEC (New Engineering Contract) (Fellows, 1995; Murdoch and Hughes, 2000). In this case, the client took its time to decide and to ask around, without communicating 'officially' on a regular basis with the tender participants. So it was in fact quite a 'loose' situation, which could function well in several circumstances, but could also result in quite a negative or unsatisfactory situation, at least for the tender participants. Nevertheless, for the client this might still work positively, because such an approach can keep the 'playing ground' as big as possible. However, despite the potential positive or negative implications as mentioned above, the participants agreed to work with this client in this way, because they did not want to lose the existing good contacts with them for such new projects. This was in fact a quite logical behaviour for them, because following existing clients towards their other new markets is a well-known strategy for building a business in a new market, thus following existing personal relationships (Tijhuis, 2002).

Such situations often happen, where companies try to acquire as many new business opportunities as possible. Especially in markets characterized by strong collectivism, the maintenance of good relationships with (local) parties is essential, because there disappointing one means more or less disappointing all. Of course there are differences regarding such characteristics for certain individuals, but, for example, Hofstede's research described such characteristics in general for the neighbouring country Iran (Hofstede, 1980). There, this collectivism was especially combined with a high level of uncertainty reduction, this possibly being a reason that a 'local for local' approach was still quite often used. Although Hofstede did not include Turkey in his research, these characteristics may also be representative of Turkey and a considerable part of the Middle East. However, as already mentioned for Turkey, it is especially interesting to see that the country seems to be more or less functioning as the bridge between the Western European Christian cultures and the Asian and Middle Eastern Islamic cultures, and also has a growing young population.

112 *The practice: international case studies*

Looking in more detail at the negotiations within the total tender process, as part of the total construction process, further progress in the situation was made by the participants, which formed their bidding teams as members of the participating consortium as follows.

1. The Belgian business partner with specialisms in industrial facilities for this client selected a Dutch developer, active and experienced in this region. Together they joined and formed a development team, in which the Belgian business partner was the contact-line to the client.
2. This development team joined with a Turkish strategic investor, which was experienced in, for example, sale–lease-back situations in (industrial) real estate. The Dutch developer coordinated the contacts with this investor.
3. The Turkish investor suggested selecting a specialist Turkish contractor, which was experienced and with which it had had good experiences in the past. Contacts were via the Turkish investor and the Dutch developer.

In the meantime the client had appointed its British architect to coordinate the preparations for the new design and the start-up of production, after completion of the plant. That seemed quite early in the process (i.e. the business partner for delivering the planned production facility had not even been selected), but there was not a long time-span available, which was another very influential factor.

- The client had calculated and requested a time-frame of about one year, in which building permission, work preparation and realization had to be arranged. So there was a quite tight time-frame for such a complex project.

However, what became clear to the consortium during the total process was the following important detail:

- The client and its architect had obviously already been searching for a possible business partner for about one year, meeting with several other, local parties, negotiating about pricing, and had also already visited several proposed possible sites in the region during that period; so the newly selected consortium were more or lesss functioning as their last hope.

After the architect had provided the schematic design and programme of requirements, the consortium started to investigate the requested details, materials and so on, and the contractor started to design rationalized proposals as alternatives to the provided schematic design. A selection and detailing of more or less comparable construction works from their earlier reference examples were used to propose proven alternatives for the client, and thus used in the final proposal, made by the consortium.

The practice: international case studies   113

Parallel with the above technical activities, the search for a suitable location to realize the requested project was started. In this case it turned out that the Turkish investor was already the owner of a possible site, but this site was not very suitable because of its location: it could be reached only via a quite narrow and steep road (so a location in a real 'greenfield' situation). Therefore other sites were selected as alternatives, of which one had quite good characteristics: well connected to the highway, with a flat surface and good and easy soil conditions to realize the planned construction works. Last but not least, the price of the site and its necessary preparation works (levelling, foundations, etc.) was reasonable, too.

The consortium decided to offer the rationalized design, completed with proven technical details, and intended to locate on the selected site. However, to be sure about the planned alternative design and the site, the consortium contacted the client through the Belgian business partner, which organized a meeting in Turkey to visit their existing operations, with the following goals:

1   The main goal of the consortium was to have enough time to discuss in further detail, and also to visit the intended site together.
2   A parallel goal of the Belgian and Dutch partners especially was to see how the parties involved would react within such a multi-national group of stakeholders; in particular, how the Turkish parties would work together, that is the Turkish representative (local operations manager) of the client vis-à-vis the Turkish investor and contractor.

This led to a very interesting meeting, as follows.

The consortium arranged to visit the client in Turkey; however, the contractor could not attend because of its tight time-schedule. Nevertheless, instead the client agreed to join the meeting during the planned subsequent visit to the site. At the client's local office, the first handshakes were really good, but from that point on the meeting soon became a little stressed. The following happened between the parties involved.

As the Turkish investor had already discovered during the few days before the meeting, it was suddenly clear that the project had been 'on the market' for quite a long time; at least, the client had obviously been speaking with several other consortia before, which made the atmosphere slightly negative. This issue had obviously not been clear beforehand, because the Belgian and Dutch parties did not have their own local 'antennas' to hear exactly what was going on there. However, during the meeting this issue was quickly mentioned within the first five minutes by the Turkish investor, but this was experienced as quite an insult to the client's Turkish representative. After this, the atmosphere resulted more or less in a situation in which the Turkish parties tried to impress and overrule each other with experiences, references, knowledge and so on.

However, the Belgian and Dutch parties could not really intervene effectively, because this seemed to have become a clash between Turkish sensibilities. Nevertheless, at one moment after about 15 minutes, the British architect evidently grew bored with the discussion; he intervened with a very short and strict injunction: 'Do not waste our time; we need a serious solution!' After all, one can understand his quite rude reaction, because he had already been waiting so long for a real solution during his previous meetings with his client and other members of the consortium during the past year.

Nevertheless, the total atmosphere stayed quite negative, so the Belgian and Dutch parties could not do much other than to propose to end discussions for a while, and prepare to leave for a joint visit to the site. Everyone accepted, so the journey to the site started; the investor took the initiative by driving in front, together with the Belgian and Dutch business partners, and the client with his architect following in a second car. As agreed, the contractor was waiting for them on the site. However, on their arrival, the client was not very enthusiastic about it, but agreed to wait during the weeks after for a serious proposal, based on that selected site. This was agreed jointly, but what did the Turkish investor do during that discussion on the site? He suddenly proposed that the client visit his own 'greenfield' site as well, to see it and compare it as a 'less possible' alternative. So they did, but with a quite unsatisfactory result: during the drive to that site, they had to contend with quite difficult road conditions, steep hills and so on. In short, the client and his architect really experienced a mismatch of the location for that site, which they expressed to the investor at their joint arrival. The investor agreed and confirmed this, but said that he had felt that they had to visit this one too; then the British architect became really angry, saying that this was 'really a waste of time'. This resulted in the initial Belgian business partner having to put in some extra effort to try to calm the client down in this situation; and he succeeded.

Figure 4.13 is a view of one of the visited alternative possible building sites, with its quite difficult road and slope conditions.

Altogether, the final outcome of this meeting was thus a quite negative atmosphere between, on one hand, the client with his architect and, on the other hand, the consortium. The experience was made all the more negative by the additional visit to the other site.

However, importantly the client confirmed that he still wanted the consortium to prepare a serious bid, based on the primary selected site; so it was prepared.

After a few weeks of preparation time, during the next round of discussions, it became clear that the client had suddenly opted for another strategy. Instead of leasing of a new foreseen production facility, it had decided to merge with a local competing producer, and thus to use the competitor's existing production facilities. The market situation had necessitated this, the client explained. This meant that new production facilities were no longer needed, and so the consortium was disbanded for this project. This was a typical experience, with no opportunity for the consortium to receive compensation for the initial transaction costs from the client.

Figure 4.13 An alternative building site.

Taking an overview of the case as a whole, the professional parties involved in the project represented in general the following behaviour.

- **Developer**: Because of the international relationship between the Belgian partner and the North American client, the coordinating of the mutual interests with the client was obviously the most important task for the Belgian partner. The Dutch partner was the coordinator of the relationships with the Turkish parties (i.e. investor and contractor). Altogether, the developers had quite a complicated balancing task, because they felt that the local Turkish parties always would have an advantage related to in-depth market influence and so on, which incorporated the latent risk that the Turkish parties would 'walk away' with the client, despite of all kind of signed non-disclosure agreements (NDAs), contracts and the like.
- **Contractor**: It acted quite passively, although it was really inventive during the rationalization of the available schematic design and search for a suitable location other than the investor's initial 'greenfield-site'.
- **Investor**: The developer asked the investor to join in him in the deal because of his long years of experience and good reputation in the Turkish market. However, that quite strong reputation on a regional and national level obviously also led to his acting as he did. Regarding the

client's impressions, he showed a much too forceful attitude, not acting exactly in accordance with the agreed consortium guidelines, but merely following his own agenda.
- **Architect**: Although the architect did not disclose beforehand the history of the search for a suitable location and partners for delivering the planned facility, his nuisance value in this obviously too long process was shown very clearly during the discussions: reacting very sensitively, he was hardly able to discuss things on a supportive basis, but was much more driven by a negative sentiment regarding this total approach, especially because he felt the stress of a tight time-schedule for reaching a feasible and satisying solution.
- **Client**: Although the North American company was in fact the decision-maker because of its shareholder role, the Turkish representative carried out the negotiations. He was also quite sensitive regarding the behaviour, especially that shown by the Turkish investor. Obviously that did not help to build mutual trust between the client and the consortium.

Gradually it became clear that the behaviours of specifically three main parties were the main reasons for disturbing the construction process.

- **Investor**: Although he was part of the consortium preparing a joint bid for the planned production facility, it was obviously difficult for him to keep following the jointly agreed approach during the negotiations. The quite forceful handling and self-presentation shown by the investor negatively influenced especially the process of building trust between the client, its architect and the consortium, which delayed the total negotiations as part of the total process. In the end, the consortium was not succesful at all: the client adopted a strategy of joining forces with another company with production facilities in the region, instead of focusing on a new production facility for itself alone. By then, the project opportunity was over.

    Comment: As the Belgian and Dutch developers were quite dependent on the reputations of the North American client on one hand and the Turkish investor on the other hand, it resulted in a quite difficult balancing act of mutual interests. In fact, initially the Turkish investor (and also the Turkish contractor) could be characterized as a kind of 'third party', because neither of them was involved in the initial contacts with the client. However, the risk of joining with a Turkish investor/contractor for this Turkish project, managed by a Turkish representative of the client, resulted in a quite complicated situation: how to keep the Belgian and Dutch developers' influence 'on board'? That was what bothered them, and that feeling was even strengthened by the way the negotiations went. The properly signed NDAs and contracts were not enough to keep the consortium jointly following the agreed strategy during the negotiations with the client. The Turkish parties simply did their own thing (obviously following their private agendas) and displayed a rather forceful attitude, which was more or less experienced by the client as a kind of arrogance (with all its negative effects on the relationship with the client). So, ultimately, this was quite a typical example of how to build (or rather destroy) trust between parties (Laan, 2008). However, by reviewing continuously the risks during the negotiations, trust could still have been kept well balanced; but especially within a consortium of several parties this requires a well-coordinated approach and behaviour of all parties involved instead of 'random' and/or 'hit-and-run' actions by individual parties within such a consortium.

- **Architect**: During the negotiations he continuously acted on the basis of a negative feeling, because of the delay he had already experienced during earlier negotiations with other parties; however, that fact was explained to the consortium only during the negotiations, instead of at the beginning. The resulting time-pressure he felt to complete his task will have influenced him heavily regarding the way he acted, which was experienced as quite a reason for delay during the negotiations.

  Comment: As obviously the consortium was not informed completely about the previous negotiations with other parties, the consortium was not aware of the strong time-pressure under which the architect and the client were working; commercially this was of course understandable from the viewpoint of the architect and his client, as he did not want to lose his negotiating position by telling the consortium that he was under time-pressure; that would have strengthened the feeling of the consortium that they were 'the only hope' for the client in the tender procedure. Nevertheless, the architect's behaviour was based on a kind of negative attitude, obviously not expecting a positive outcome at all regarding the negotiations with this consortium. That negative attitude was fuelled all the more by the arrogance shown by the Turkish investor towards the client and his architect. Although one can say that the behaviour of the investor should not have been like this, or was an example of the typical 'masculine society' of Turkey (Hofstede, 1980, 1988), still another issue is that the architect could indeed have been somewhat more positive and open about discussing a feasible proposal by the consortium, instead of only acting from a negative viewpoint, driven by his past experiences with other parties and time-pressure. Nevertheless, in this case one can say that the process of establishing (organizational) relationships (Ring and Van de Ven, 1994) was indeed disturbed; but here it was caused more or less on both sides of the negotiating table.

- **Client**: As the Turkish operational manager, representing the North American client, was fully aware of the strong reputation of the Turkish investor, the client developed quite a sense of alertness, somehow scared of being 'overruled' by that investor. This was obviously more or less driven by a latent fear, fuelled by an architect driven by time-pressure,

118 *The practice: international case studies*

confronted with an investor, driven by power. Altogether it resulted in quite chaotic negotiations on both sides of the negotiating table.

Comment: The fact that the client did not really act according to a strict procedure, combined with the experienced time-pressure and the latent fear of the investor, created a kind of chaos or vacuum, in which the attitude of especially the local parties (investor and contractor) led them to behave according to their own devices; this despite the NDAs, contracts and so on. So in general one can say that, within this kind of (tender) process and negotiations, there is a strong need to structure the process, especially if one wants to create and to keep a level playing field. Such balances are of course regulating not only the procedure itelf, but also the way to behave, how to judge the outcomes and so on. In fact, it will add as a tool for balancing flexibility against efficiency, which is a theme that often occurs also in other branches of industry, and also on various decision-making levels (Blanken, 2008). Nevertheless, a thorough briefing and debriefing of all the parties involved during such negotiation processes is vitally necessary, especially for keeping the mutual interests well balanced to reach a feasible and satisfactory solution.

A final circumstance during this case was the fact that the client changed its overall expansion strategy on the grounds of market circumstances, and decided to join and integrate its foreseen expanding production capacity within the existing production capacity of a competitor in the market. Therefore there was no longer any need for a new production facility.

Based on the 3C-Model™, one can represent the above situation as 'types of behaviour', being:

- positive (i.e. speeding up the running process);
- negative (i.e. slowing down the running process).

Although in this case the project did not lead to a contract at all because of the changing overall expansion strategy of the client, it still contained a conflict situation. During the process of moving forward from contact phase towards the conflict phase, there were changing attitudes of the parties involved; here especially the architect and investor changed their behaviour, being more or less influenced by their mutual behaviour based on time-pressure (architect) and power-show (investor). Already during the first contact phase this led into a negative disturbing manner during the discussion about the proposed technology solutions. Finding a satisfactory solution to maintain progress became quite difficult then. In Figure 4.14 this is represented schematically in the three phases of the 3C-Model™, which model was also referred to in another research by Turkish parties regarding the existing business culture in the Turkish construction industry (Akiner, 2005).

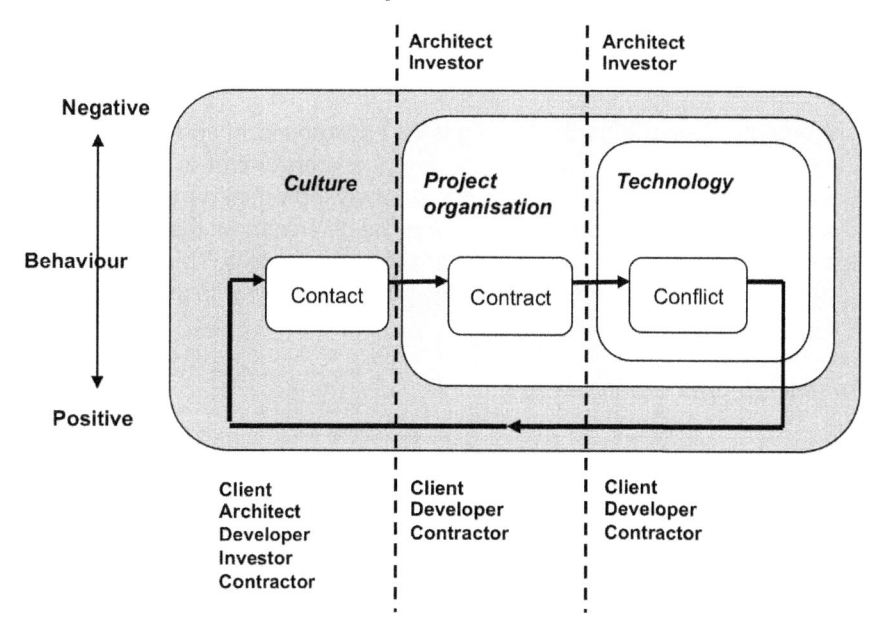

Figure 4.14 Types of behaviour (positive and negative) of the parties involved in the three phases of the 3C-Model™.

Apart from the fact that the project negotiations were not followed by operational contracts because of the changing overall expansion strategy of the client, the process itself was also a quite unsatisfactory project for the consortium. A stronger focus on keeping the consortium members on the same goal (by speaking with one voice with the client and other external parties) was from that moment one of the leading issues for the next projects to come within their project investments.

Conclusions and recommendations: lessons learned

Contact (culture)

LESSON 1

If one is selecting parties for a possible joint project, and the planned consortium members are of several nationalities with no joint track-record in previous projects, there is a large risk of unmatching behaviour. How to keep all the parties involved in the consortium acting as one? This is true not only if working with several nationalities, but also and in particular if consortium members are of the same nationality. This does not guarantee a smooth negotiation process at all, because the same (group of) nationalities within a consortium can be the reason for uncontrolled following of (joint) hidden agendas. So be aware of the differences within those groups of nationalities, which may be somewhat unclear to foreigners, but which are still influencing their behaviour. Therefore, lesson 1 is defined as:

> Although parties do have their national background in common, their behaviour within a multi-national group or consortium is not automatically an equal one; this is because every individual party will work according to its own agenda, which may be different from the other parties involved, regardless their own or others' nationality. Therefore, the matching of team members should be done especially with the focus on matching the goals and agendas.

Contract (project organization)

LESSON 2

Receiving an invitation from a possible client to prepare for a bid on a foreseen contract is always an interesting issue: one is on the one hand honoured by this invitation, and on the other hand challenged to prepare a competitive bid. However, if the procedure under which such bids are arranged by the participants and judged by the client is not fully clear, one should be warned to keep away from uncontrolled negotiation processes. Therefore, lesson 2 is defined as:

> The procedures used by clients for tendering for specific contracts should be fully clear to parties before they are invited to join such tender procedures. This to avoid unclarity because of uncontrolled negotiation processes, which will not lead to a satisfactory result for all the parties involved, but will instead lead to a waste of time.

Conflict (technology)

LESSON 3

Preventing possible conflicts is always a hard issue, especially because one can be surprised by unexpected behaviour on the part of parties involved, which may lead to an unexpected reaction by the other parties involved. Therefore a thorough estimation of how parties will react in certain circumstances can help others to avoid getting into conflict with them. An important issue is then the investigation of the history and/or background of comparable project initiatives and parties involved: How were situations handled previously? What happened before? Why are potential clients approaching us in this case? Under what circumstances are our existing clients working? What is influencing their behaviour? and so on. Therefore, lesson 3 is defined as:

> For solving conflicts it is important to focus on the present situation and future results of actual decisions to be taken. However, these decisions

are often perceived by other parties involved based on their previous experiences during comparable situations. Knowing their behaviour during such comparable situations and/or projects within their background can help to keep the conflict situation under control, and in working towards a feasible solution. However, one should always be aware of (unexpected) actual circumstances, influencing the behaviour of the parties involved, making the prediction of such behaviour quite difficult and complex.

Resumé

Following existing clients into new markets is always an interesting strategy in growing a business. However, one should be still aware that those existing clients may show different behaviour if confronted with other parties, working in such new markets. This can lead to a need to continuously review the playing field and all the parties involved, thus keeping alert to the behaviour and interests of the existing business partners as well as those of the newly involved business partners. They are all influenced in one way or another by especially unexpected actual circumstances, which fuels the need for being alert to them as much as possible.

Case study 5: Designing a production factory

The following case study describes the design process of a production factory in one of the booming United Arab Emirates. Although the Middle East has recently undergone substantial development, with several large initiatives resulting in masterpieces of newly designed projects, it was not always thus. Before the discovery of oil and gas after the Second World War, this region was a desolate environment, largely desert. Only the local tribal communities lived in the desert, and there were hardly any foreigners. However, after the discovery of oil and gas, this changed dramatically. In, for example, Saudi Arabia the huge oil and gas revenues fuelled the economy, so that a strong growth of investments started not only within their own country, but also through their investment institutions elsewhere in the international marketplace. Several nationalities therefore entered these regions, especially driven by the huge investments in the oil and gas industry, so that gradually these regions became increasingly wealthy, with a large number of expats working there, also introducing typical expat issues related to managing dual worlds (Black and Gregersen, 1992). Such developments did not occur only in Saudi Arabia: neighbouring countries such as Bahrain, Kuwait, Qatar, Yemen, Sudan, Iran and Iraq also profited from their natural energy resources. Although following the Second World War the region was greatly affected by several other wars (e.g. the war in the Lebanon and the first and second Gulf Wars), (re)developments then and now prove that those former war zones can still recover and find a path towards strong economic (re)development. Consider, for example, the case of Lebanon: after its war during the 1970s and 1980s, the country has found a growing strength to become again one of the hotspots in the region (Republic of Lebanon, 1998).

Although such developments are on the one hand often driven by commercial reasons, post-war regions there and elsewhere will on the other hand also profit gradually from (funded) development programmes by their governments or by other (international) organizations; not least because there is an increasing global awareness of the need to pay attention to post-disaster reconstruction, whether following wars or natural disasters (Lloyd-Jones, 2006). However, in all cases there is still the need for humanitarian relief, by (re)building the future of those regions. So indeed this is an increasing challenge for the construction industry as a whole, to take up this task by turning it into a business opportunity.

Project description

The design project that is the subject of this case study was located in the new industrial zone of one of the United Arab Emirates (UAE). The case study describes a typical situation, as the design period (*c*. 2006–2007) coincided with a huge construction boom, particularly of offices and housing/apartments. However, these projects were located along the shoreline, whereas the site available for the new production facility was located inland, in fact on the edge of the desert area. In general, this project can be described as follows.

The leading client was an Indian company, active in the international energy business, which had established a joint venture with a local Middle Eastern company. This joint venture was in fact the client, for which the Indian business partner held the responsibility for the overall coordination of the design phase until the final realization phase. Because the joint venture was already rolling out a growing business in the Middle East, it had decided to build a new production facility instead of just importing goods from the Indian factory. For this reason it was starting to investigate the needs, thus working on the programme of requirements. In general this programme of requirements was listed as follows:

- total ownership of the site and the facilities (available area *c*. 50,000 m^2);
- optimized production logistics needed in the total design concept;
- representative design and style, with a proven concept, if possible;
- competitive price/quality level;
- if needed, international parties involved to obtain the best results.

In general this was a quite basic basic programme of requirements for the design of the planned production facility, with obviously a great deal of freedom for the designer regarding shaping, style and so on. However, for the construction itself the client was still focusing on local parties, because its approach was to try to find the best proven design solutions if possible,

The practice: international case studies 123

but it was planning to choose the local contractors to turn these proposed solutions into reality. Therefore:

- construction itself to be organized/contracted by the client by his local coordinating engineering office.

In this project it was clear that the client had choosen a fragmented approach, instead of an integrated (design/engineering/construction) approach. This fragmented approach was in this project not a real problem, because the client knew what it was doing: selecting the best possible solutions and their involved parties, combined with keeping control itself of every step in the total construction process. However, rather than this chosen fragmented process approach, trends in international business are increasingly going in the direction of an integrated process approach, supported by several innovative IT developments, driven by platform-driven tools with or without internet-based utilities, but all based on improved collaboration systems (Mulholland and Earle, 2008; Fingar, 2009). Recent leading publications regarding business operations improvement describe these kinds of developments as 'The New Paradigm in Enterprise IT' (Baan, 2010). Nevertheless, as experienced often, construction industry is still not really following this trend generally, so there is still a long way to go.

Apart from the fragmented process approach, another thing was that the client paid special attention to the selection of the architect, because the final foreseen design and style was an important issue for the the client. This selection of the architect consisted of the following phases, not really organized in a structure but merely a combination of circumstances.

1. It all started with the primary selection of suppliers for the delivery of the foreseen special production machinery; several possible suppliers were selected and visited.
2. During a visit to one of those suppliers, located in Europe, the client was quite impressed by a production factory which was located nearby.
3. Following investigation, the supplier was able to tell the client that the factory had been designed by a specialist Dutch architect. The supplier also recommended the architect as a reliable and creative business partner. In addition, the client viewed some of the architect's portofolio on the internet.
4. Therefore the client had preselected this architect to request to make an offer for the total design, searched for his contact details, and tried to contact him directly with a telephone call; however, the architect did not react at all but left the message on his desk.

Although the described selection process worked quite primitively, it still contained two of the essential elements needed for effective communication:

124 *The practice: international case studies*

- information;
- contacts.

However, what the architect was missing was an 'antenna' for multi-national project opportunities, involving multi-cultural ways of working. In fact, the architect did not take the contact seriously, because the firm often received vague marketing and project requests by phone and email from international contacts. These often proved to be 'spam' sent to unknown users from unverified contacts or containing marketing information that violated international regulations (Venekatte, 2002). Thus, in this case also, the architect assumed that the initial contact email was just spam.

So it was more or less a coincidence that the architect eventually responded positively to the client's repeated requests. Later in this case study this contact process will be described in more detail.

Figure 4.15 was taken during preliminary investigations and shows the Dutch architect and his accompanying consultant at the anticipated location, a large area of desert in which four flags delimited the proposed building site.

Building in this desert region was a quite difficult situation, not least because of the extreme circumstances regarding techology. For example, using concrete on site is quite difficult because of the extreme local temperature (higher than 50°C in summer), which introduces the latent need

Figure 4.15 Investigating the proposed location.

to cool down the concrete. Using prefabricated structures with production in the factory under controlled conditions makes this much easier (Zein Al-Abideen, 1998; Odijk, 2010). Another example was the influence of multi-national parties involved in the construction process; although this case study focuses on experiences during the design phase, the setting in the Middle East forced the parties involved to act according to the specific local circumstances (Enshassi and Burgess, 1991). These circumstances were not particularly easy, as will be described during the following analysis.

Stakeholders involved

In this design project, several stakeholders were involved. The following stakeholders are mentioned, including their roles:

- **Client**: a professional joint venture between an investment company from the Middle East region and a production company from India. The Indian company was especially responsible for the daily negotiations and overall coordination, whereas the Arabian company was primarily responsible for the financial issues and local registrations.
- **Engineer**: as a professional company from the Middle East region, it was responsible for the structural engineering including the preparation of the planned working drawings and coordination of the contractors on site.
- **Architect**: a professional Dutch party, and a creative architect but with a low level of experience in international projects. It had never worked before in the United Arab Emirates.
- **Consultant**: as a professional party with international departments, also in the Netherlands, it had experience in international developments and construction processes, and had also prior contacts in the UAE.

Differences in behaviour of the stakeholders involved

During the project, it became clear that lack of several business attitudes, as representative of differences between the business cultures involved, can make or break a project. Nevertheless, the start was an interesting event, because the client showed a proactive attitude by trying to contact the architect directly by telephone. However, as mentioned above, the architect did not really take this effort seriously and did not react. However, the following happened in this case.

At a reception attended by the architect (who was the manager of the architectural firm) and some of his employees, they met one of the managing partners of the consultant firm. Although it was a purely general conversation, the discussion came gradually around to the topic of international business and its impact on the construction industry. This was (CIOB and

126 *The practice: international case studies*

CIB, 1987) and still is (Langford and Hughes, 2009) an issue of interest, dealt with in several publications and conference themes.

Following the discussion, the consultant asked if the architect had ongoing international design projects at that moment; the architect first anwered negatively, but then reminded himself of the still unanswered repeated request from the Indian representative of the possible client.

The consultant was surprised: Why not react seriously to such a request? Did the architect realize what opportunity he might be missing? And so on. The architect still was not convinced that it was worth taking action, but because of the good long-established relationship between them they agreed to arrange as soon as possible to check the files so far.

And so it happened: the next week they had a meeting and discussed the pros and cons of reacting positively to the request. In general this meant checking the possibility of a design deal according to the following criteria from the architect:

- The client should be a trusted party, with a sound financial background and payment attitude.
- The legal situation should be as clear as possible; especially keeping the Dutch, internationally translated, standard regulations as a standard framework for their possible design contract.
- Design style and project location should be challenging, thus giving the opportunity to create a lighthouse product.

The consultant added the following criteria:

- Total coordination should be attempted; at least the whole design process should be organized under a non-fragmented approach, in which the consultant would take the lead in the overall coordination.
- If the Dutch standard design regulations were not accepted as guidelines, other international design regulations which were comparable to these should be chosen.

At first sight, the most important were the financial and legal criteria, especially because the architect and his office had never worked in the UAE. Therefore, the managing architect had agreed to work according to a certain plan with three main steps, which was explained as follows.

- First step: Contact the client, and discuss with them the ins and outs regarding their plan for a new production factory. Although not physically meeting then, the telephone calls and email correspondence should of course be as clear and transparent as possible, to get the right impression of wishes and expectations.

The practice: international case studies 127

- Second step: Discuss the planned work package, its process implications, such as planning and scheduling, and the legal framework. At least during this step, it would be necessary to meet personally to discuss in more detail and foresee if the negotiations could lead to a serious deal.
- Third step: Altogether, the contacts and negotiations about specifications should lead to the preparation of a serious formal offer for the requested design activity, ultimately resulting in a possible contract. Only after the signing of the contract should the real work be started.

In fact these three steps were comparable with the described 3C-Model™, especially because the business culture would play quite an important role in managing this international process, leading to the foreseen design deal (Tijhuis, 2005). In general, this meant that the architect and the consultant agreed that the following negotiations had to be done in line with two main goals in mind, as follows:

1 to clarify the business issues and get the contractual arrangements secured;
2 to realize a sophisticated and professional 'lighthouse' design project.

However, it was pointed out that keeping this sequence would become a very important success factor for this deal.

What happened? As the architect was convinced by the consultant that the opportunity could indeed lead to a serious deal, they agreed that the initial contact messages should be used to prepare a further detailed response by email. The architect appointed one of his employees to coordinate this, and the consultant started to prescribe this planned response. Gradually also the architect's employee became enthusiastic, so the communication process really got going. As the Indian client reacted positively now after the first email correspondence, a conference call was arranged. In the background the above two main goals (and their intended sequence) were kept more or less as a guideline for the process of trying to reach a serious design deal.

During the conference call, the general impression and atmosphere were experienced as positive by all the parties involved. So, after a few days of communication process and online exchange of information regarding reference projects, foreseen basic programme of requirements and so on, the need emerged to arrange a visit to the client's representative office in the UAE, together with a first visit to the site. That would be a welcome opportunity to meet them, because the client's organization was a joint venture between the Indian client and a Middle Eastern organization.

Flight arrangements were coordinated by the architect's office, but then it pointed out that the consultant's agenda did not fit into the planned itinerary. So, instead of adapting the schedule, the architect accepted the client's suggestion to go ahead with the travel arranged, but then without the consultant attending; in this case only the architect's responsible design

128 The practice: international case studies

employee would go, and this would not be a problem according to the client and architect, because it was just a first meeting. The consultant agreed, and reminded them to pay attention to the issue of not accepting any work unless at least a basic contract was signed. So the employee went to the UAE, meeting the client for a first impression and fact-finding-mission.

It was indeed an interesting meeting because, after returning, the architect's employee mentioned how impressed he was about the local situation, the willingness of the client to choose them for designing the new production factory and so on. In addition to this, the client had already asked for additional information about some thoughts or suggestions for certain design solutions to be sent to them by email. The employee had agreed to this request, simply to show goodwill.

In the meantime the consultant and the architect, together with his employee, were starting to prepare a model contract, on the basis of which the design work could be executed. In fact the architect insisted on using the Dutch nationally used standardized model contract 'DNR 2005' (*The New Rules 2005: Explanatory Notes on the Legal Relationship Client–Architect, Engineer and Consultant DNR 2005 – Standard Form of Basic Contract*; BNA and ONRI, 2005). Although this is a Dutch basic model, it still has functional aspects for international projects. So, together with the consultant, the use of the Dutch contract model was communicated to the client, and he agreed to this. This seemed to be an extra indication that he was really willing to go into business with the Dutch architect, especially because international parties from countries such as India (as part of the Commonwealth) often base their design contracts only on, for example, British-originated architectural contract models from RIBA and the like (RIBA, 1996a,b).

Additional information about insurance, finance risks and so on was obtained through specialized partners in the Netherlands, and within this intended contract framework it was estimated that there was no high level of risk to contend with, so the negotiation process was ongoing.

However, before signing any contract, of course a well-structured estimate of the costs of design work was needed. For this, additional information from the client regarding estimated scheduling, requested level of design details and so on were asked and delivered, so the basic outline of the initial offer for the planned design works was prepared by the architect in collaboration with the consultant. After this had been communicated to the client, new flight arrangements were made for a more detailed discussion, and the architect's employee and consultant prepared for the next visit to the UAE. However, what the consultant did not know at that time was that the architect and his employee were still sending examples of additional design information to the client on his request, merely thinking to show their willingness to working on a collaborative basis. However, the design contract was still not signed.

During the second visit to the UAE, and during the discussion with the client, the consultant was confronted with the fact that the architect's office had already sent (a considerable amount of) design information to the client by email, which could be used by the client and his local team to compose the desired design. This was an initial risk to the final closing of the deal, especially because during this visit the client's representative was also attending one of the meetings. That representative was an engineer and the manager of a Middle Eastern engineering office, being responsible for the final working drawings and coordination for obtaining the necessary building legislation. Although the meeting was good and the atmosphere pleasant, it still had a kind of uncomfortable feeling. Would the client indeed be willing to sign a contract? What would he do with the information already given to him by the architect's office? At the end of the joint meeting this feeling was intensified, because the engineer showed a kind of simplified design, in which he had integrated several of the design principles provided by the architect, so this issue really confronted him during that visit.

However, after that meeting there was a visit to the central board of the industrial park in which the intended building site was located. The experience was interesting for all parties involved, especially when the architect's employee asked the central board of the industrial park if any subsidies were available, for example if solar cells were to be installed on the large roof surfaces, a quite logical suggestion considering the huge amount of sunshine during the year. However, the quite amazing answer was that it would not be useful, because such solar energy was very difficult to keep under optimal operation, because the solar cells would often be damaged by the blowing winds and grinding sand-storms. In addition to this they remarked that water was the real problem, not the lack of energy, because of their large national energy reserves; therefore the desert area of the site had to be designed with as many plants, trees and so on as possible. This was another quite contentious issue during that visit, also more or less proving the fact that local situations and (business) culture do indeed influence the decision-making processes and (business) policies (Beamish, 1996; Tijhuis, 2001b).

After completing the business on the agenda, the architect's employee and consultant joined the client at his hotel for the closing meeting and dinner. It was a good meeting, during which several pieces of information were exchanged in a quite informal setting. Over dinner, several aspects of the clients' goals with this project were discussed. The main point was that the Middle Eastern partner had still to make a final decision about the financial aspects of the total design offer. In parallel, although the client still had decision-making power over this project and thus also for the design work and expected contracts, he put the existing family culture first, with the following result. He excused himself by the fact that he had to discuss the design offer in more detail with his family back home in India, especially as they were also involved in this business, but merely responsible for the Indian business of the company. So obviously also according to the Indian business culture, the family connections as an example of collectivism played a quite strong role in the overall decision-making-process; indeed, for Indian people this is a representative business culture and behaviour, merely based on group values and status (Negandhi and Reimann, 1972; Hofstede and Hofstede, 2005; Ray and Chakrabarti, 2006).

After returning to the Netherlands, the architect and his consultant tried to get the contract finalized, estimating in the meantime also the actual risk that the client would be able to work in more detail on the design with the assistance of his own engineering office. Therefore closing the deal as soon as possible was crucial to be able to send an invoice for the work already carried out by the architect and still unpaid. However, for the consultant there was less financial risk, because he had a direct contract with the architect and was paid on a general fee basis, independent of the design contract.

How did this part of the construction process come to an end?

1 The consultant was not able to change the sequence of activities, thus having to accept the unsatisfactory situation that the client had already received from the architect and his employee a quite good portfolio of examples and basic proposals for possible design solutions.
2 This portfolio was delivered without any signed contract between the client and the architect, thus bearing the risk that works up to that moment would not be paid for, or even that there would not come any further design works at all for the architect in this project.

This siuation was thus quite unsatisfactory, especially for the achitect's office, but he had to agree that the consultant was right when listing the special two goals and their sequence during the first internal briefing before accepting this opportunity. Remarkably, however, the client still asked for additional information and help for the final design scheme, while the discussion about the concept design contract was still ongoing, especially about pricing levels. This resulted in a quite difficult situation for the architect:

- Accept this situation, merely to keep the goodwill, and try to save as much as possible from the expected design contract?
- Decide to quit and lose any chance of payment?

Together with the consultant, the architect decided to try to go for the 'golden mean', more or less following an attitude, often seen as a typical business culture of Dutch business parties (Hofstede, 1980; Hofstede and Hofstede, 2005), which is generally referred to in the Netherlands as the 'polder model': for example, trying to create a state of 'symbiosis' between parties during negotiations instead of just following a kind of power play; however, according to such approach the generally foreseen goal would then merely be a final deal with a certain mutual profit. In the end the client agreed on this, and the architect accepted to finalize only this phase of the preliminary design including basic details and design principles, while the Middle Eastern engineering office would then finalize the design on behalf of the client, and would prepare for building legislation procedures, working drawings for the planned contractors and so on. So after all, the compensating invoices from the architect were agreed and paid and the (relatively short) design phase for the architect was finished. However, the atmosphere between parties was kept as positive as possible, because according to the parties involved there were 'no hard feelings' after all.

Considering the total case, the professional parties involved in the project represented in general the following behaviour.

- **Client**: Although this was a professional joint venture, it was shown that the differences in behaviour between the Indian and Middle Eastern partners were quite evident. The Indian partner was representing the joint venture, but still had to ask internally regarding financial issues; parallel to this, the Indian partner also relied heavily on the initial family structure of the business, originated in India. However, it was obviously trying to postpone the moment of signing a design contract with the architect as much as possible to the back-end of the total construction process.
- **Engineer**: He was just acting on behalf of the client, but proved to be very creative in the area of expanding the design on the basis of the basic design information provided by the architect.
- **Architect**: Being a professional party, the office was known as very experienced in the creative process. However, the professional drive to support the client with as much initial information as possible turned into a quite non-commercial situation: they 'forgot' to have the design contract signed before getting into too much detail in providing design information to the client.
- **Consultant**: As he knew the architect from earlier projects, he was convinced that they were able to handle this type of project. However, to keep the commercial business aspect as well balanced as possible, he added the extra common goals and their sequence, before the negotiations should start: first close the deal, then start executing the design work. Nevertheless, because he was not updated regularly about the design communication between the architect's office and the client, this resulted in quite an unsatisfactory situation of 'uncontrolled' negotiations as part of the total construction process.

Gradually it became clear that the behaviour of specifically three main parties was the main reason for disturbing the construction process.

- **Client**: Knowing how to handle this kind of situation, the client's attitude was obviously driven by the desire not to get a contract as fast as possible, but to obtain as much information as possible, thus being able to control the situation. Especially during communication, it was never shown clearly to the architect and consultant how decision-making processes were organized, which made it possible for the client to 'play' with the parties involved: he more or less explained during the meetings that he also had to ask the joint venture partner and his family how to decide about the next steps, but the joint venture partner and family did not actively participate in the negotiations at all.

Comment: Working with professionals in the business, it is always necessary to know who is responsible for what; is one really speaking with decision-makers or are they just representatives? However, as in this case the joint venture partner had decision-making power to decide about the contracts, it was known that he still had to discuss with his other partner about financials. That was clear from the beginning; nevertheless, the fact that he also had to consult his family back home in India was quite a strange aspect, but still understandable regarding these kinds of business cultures. However, it was felt on the one hand by the architect and consultant that the client should have mentioned this fact to them somewhat earlier in the initial contact phase. On the other hand, the architect, or at least his consultant, should have asked more explicitly about these issues, before starting the negotiation process.

- **Architect**: Having been quite uninterested in the initial phase of the contact, he gradually got really enthusiastic about this design opportunity. However, the professional 'design drive' seemed to overrule the awareness that closing a satisfactory deal was necessary before starting the real design activities; at least from the viewpoint of commercial business, especially in this international business setting.

    Comment: For a professional architect, a strong reputation as a real good designer is of real market value. However, especially in the international marketplace, a professional company needs more qualities, especially in the field of commercial negotiations. That was obviously just the aspect which the architect's office lacked, in general not actively working abroad regularly. As it seemed, this situation was exactly one which the client knew how to handle, showing a kind of 'charming' attitude towards the architect, thus keeping himself as client in control over the total process. The architect followed obviously quite easily the initial design requests of the client, and in the meantime did not inform the consultant about this parallel communication with the client.

- **Consultant**: He knew how to pay attention to certain international influences such as behaviour and business culture during the negotiations with such an international client. Therefore he focused especially on the specific business goals, and those in the planned sequence: first closing the deal, then executing the design work. However, communication with the client was planned that way, but he did not really calculate on the behaviour of a professional architect, which was merely following his professional eagerness about the opportunity for design work instead of first following the commercial business aspects. So this internal difference in behaviour resulted in the fact that the client was holding the lead in the total negotiation process.

Comment: Although it is a good thing that professional opportunities are followed by a certain 'drive', in modern business environments there still needs to be a kind of realism of 'getting the deal done', otherwise negotiations and other efforts may lead to nothing. This requires a strong focused attitude towards the common goals one wants to reach. However, working within a team of professionals is still a difficult issue, because every participant has his or her own ideas and perceptions, and probably also different (personal) goals. Although the consultant did foresee the risk of such different behaviour possibly occurring during the negotiations, he thought he could solve this by just defining extra goals and some simplified procedures, shown in their intended sequence. However, this turned to be not enough; extra checks and balances should have been used, or at least formal or eventually informal communication procedures should have been followed by all the parties involved to reach the foreseen outcome (i.e. a satisfactory design contract and a subsequent satisfactory exection of design work) as clearly as possible (Tijhuis and Lousberg, 1998). Nevertheless, it might not have kept this case from resulting in difficult negotiations as part of the total construction process, because the architect's main behaviour was merely focusing on keeping the goodwill of the client by seeking a kind of symbiosis between him and the client. Although using, for example, certain quality procedures or other tools during negotiation processes, it still needs a certain level of flexibility to reach satisfactory results, which is also proven by, for example, research on experiences in different purchasing processes (De Boer, 1998). However, such a symbiosis may not always function well in every (international) business setting. There are also personal and (business) cultural aspects influencing the total negotiation process.

Based on the 3C-Model™, one can represent the above situation as 'types of behaviour', being:

- positive (i.e. speeding up the running process);
- negative (i.e. slowing down the running process).

Although in this case the project did still really lead to a design contract, it was just for the preliminary design aspects, including basic details and design principles. So in fact it was quite unsatisfactory for the architect from especially a commercial viewpoint, and also quite unsatisfactory for

134 *The practice: international case studies*

the consultant from especially a process viewpoint: as the consultant had stimulated the architect's involvement, he had expected more from this project opportunity. However, because of the changing behaviour of the architect and client involved, the total process was going its own way, merely controlled by the client. Although empowering the client was and is still an actual theme in modern construction industry, the way it was done by the client did not feel good for the consultant or for the architect; the latter realized by himself that he had been too eager for the design work instead of keeping control over the commercial aspects. So after all there was a lack of well-coordinated behaviour, resulting in quite difficult negotiations, which delayed the total construction process. In Figure 4.16 this is represented schematically in the three phases of the 3C-Model™.

Although the architect and consultant felt that they did not really get enough business potential out of this deal, they were still satisfied because the client was satisfied too with the delivered design work, shown at least by his correct payments according to the agreed final design contract. So it was clear that the negotation process was not a good example for businesses to follow, but it was interesting enough for them to learn from it for other possible future opportunities.

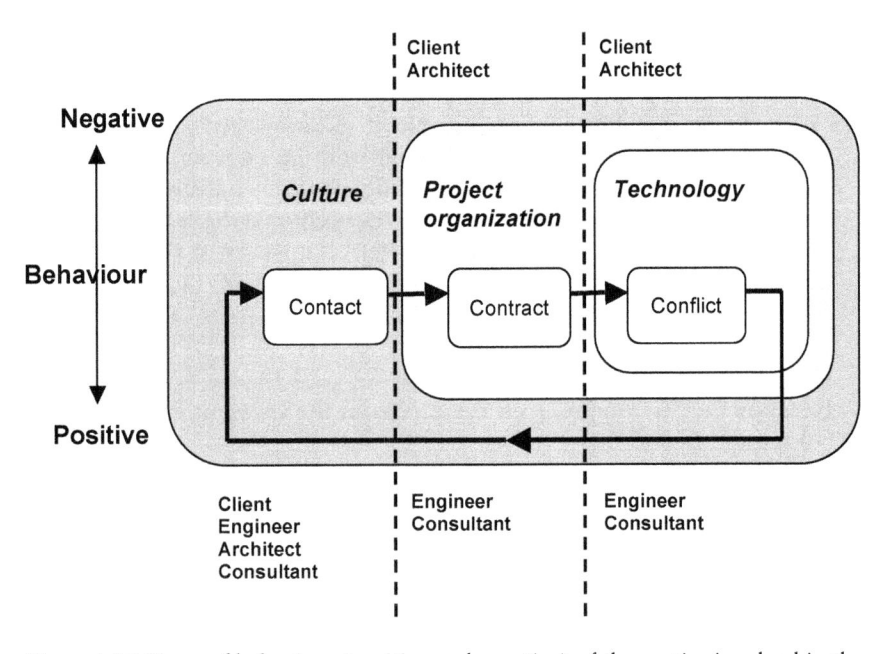

Figure 4.16 Types of behaviour (positive and negative) of the parties involved in the three phases of the 3C-Model™.

Conclusions and recommendations: lessons learned

Contact (culture)

LESSON 1

During the initial contact phase, the parties involved often show weak signals regarding their general behaviour. In this case study, for example, the architect showed inexperience in handling such international opportunities but was very eager to start the planned design work, even without a signed contract. The client, in contrast, knew exactly what to do and how to keep the involved parties interested. The consultant was probably too trustful of the architect, and believed that the architect would indeed follow the agreed sequence of steps during the negotiation process. However, the consultant was unaware that the architect was just following his own drive to make a challenging design. Such risks of deviations or unpredicted behaviour often occur when working with professional specialists or experts, who in general are aware by themselves that they are doing right in the field of their expertise. As a result of this case study, lesson 1 is therefore defined as follows:

> Working together with different specialist parties in a team needs a strong sense or awareness of the weak signals related to the behaviour of the different parties involved. In particular, awareness of the other party's professional and/or private backgrounds and drive can reduce the risk of unpredicted behaviour during the construction process.

Contract (project organization)

LESSON 2

Following the opportunity for the acquisition of an interesting professional contract always needs an important step in between, focusing on the agreement itself: on what basis should the contract be signed, with whom, about what, when and so on? This is in general a quite logical situation, but not always logical in itself: especially when a situation of mutual trust has been created, the formal signing of a contract may be postponed to a later moment in the total construction process, or even be cancelled altogether, leading to a lot of frustration by in general at least one of the delivering parties involved. Therefore as lesson 2 is defined:

> Although trusting the parties involved within their business environment can be recognized in general as a positive signal for the specific situation between these parties, it should never lead to unclear situations regarding their mutual agreement: Is there really commitment? Did

parties sign for what they expected it to be? And so on. Therefore, in business environments in general, dealing on the basis of mutual trust is good, but it should be so good that it is just logical to add the signatures to this mutual trust as well.

Conflict (technology)

LESSON 3

If in situations of conflict, for example the parties involved may want to keep the peace or goodwill, or eventually try to use the conflict as a solution for solving the misunderstanding. However, it would be a missed chance if the conflict were not used to discuss the obviously underlying problems. Although it is clear that different groups of people do use different ways of handling those conflicts, it should never lead to a lack of clarity between the conflicting parties. If one keeps acting on the basis of mutual respect, there may be some professional harm, but it should never lead to personal harm. The conflict should be seen then merely as a critical incident, which should be used positively for learning about each other and to strengthen the capacity to handle difficult situations and even prevent them from arising again in the future. Therefore as lesson 3 the following is defined:

> Within several projects the existing business relationships lead to conflicts. However, such conflicts are better viewed as an opportunity to get a better understanding of the underlying problems and clarify the situation; and this especially if seen as a critical incident, in which conflicts can even be of use for improving one's capabilities for handling these difficult situations, thus leading to better prevention of such conflicts in future. Nevertheless, especially in business situations one should always be able to separate the business conflict from the personal relationship.

Resumé

The use of opportunism in the search for new projects or clients is a daily challenge in the construction industry, and in other branches too. In general, it is part of the entrepreneurial spirit of an organization or a person. Without such opportunism the growth of a business might even become quite difficult. However, if just seen as searching for a 'lucky shot' to see whether an opportunity could be succesful or not, it might lead to a quite embarrassing situation because of frequently occurring pitfalls during such search. Therefore it seems better to follow a kind of strategy in which opportunism can still have its place, but in such a way that there are defined a set of 'antennas', which are continuously searching for weak signals about specific aspects, indicating risks, uncertainties and so on. Previous experiences related to negotiations with other (business) cultures are then of serious help for such entrepreneurial activities. However, it should be in such a way that one still can keep control as much as possible over the decision-making process.

Case study 6: Organizing an international distribution structure for special building materials

As an important part of the business in general, the portfolio of building materials and their producers and suppliers play an important role in the total construction industry. This case study describes the experiences in the process for constructing a new international distribution structure for a Chinese manufacturer of specialist building materials. China is not only a large importer of basic building materials, for example steel or cement; it is also becoming increasingly an exporter of specialist products and, for example, products for finishing works, installation components and so on. For becoming such a large 'production hub' for the internal and external marketplaces, China is not only very eager to import the basic materials, but also focusing on importing knowledge and technology; and this not only by just copying it, but also and especially by improving it and developing it into new products (Ren, 2004). Although there is still a difference in, for example, the steepness of the learning curves when transferring new technologies from technologically developed countries towards technologically developing countries (Steenhuis, 2000), it is predictable that in the near future a country such as China will take a serious place in the international marketplace for innovative building materials, not least because of its huge home market with an increasing demand for innovative, cost-saving and energy-saving solutions.

Project description

The client in this case study is a large and reputable Chinese company which has its own production factories throughout mainland China. Each develops, tests and produces a portfolio of specialist product, for example specific coatings or paints. These products are used in foundations, concrete structures, bridges and railway infrastructure, office buildings, houses and so on, so in general the products are quite uniform but variants are produced to specific recipes for specific uses. For the internal Chinese market, the client adopts the following strategy:

- delivery on demand, directly to buyers at project level;
- delivery of stock, directly to distributors with their resellers.

For the international export market the client adheres to the following strategy:

- delivery on demand, directly to existing buyers at project level.

Thus, there is a difference between the client's internal market and its export market: the export market is based on delivering at the project level to specific existing (Chinese) clients. This strategy is also called the 'piggy-back' method, focusing on 'following your existing customers'; this method is often used by business parties to enter foreign and/or new markets, and considered also as a way of reducing the risk of failure, which is high, and the associated costs (Tijhuis, 2001c). Therefore, considered from that viewpoint, the client's strategy was not strange, especially because several Chinese companies, for example contractors, are increasingly working abroad. Therefore, there were enough interesting opportunities to follow their existing or possible new (Chinese) clients abroad (Michel and Beuret, 2009).

However, considered from another viewpoint, one can wonder why such a large company did not organize its distribution channels more structurally for the international marketplace. That discussion was just the basis of the situation described in this case study.

It starts some time between 2007 and 2008, during a business meeting and dinner, somewhere in a large industrialized Chinese city. After a conference on several construction-related issues, some of the participants had dinner together afterwards. During that dinner, the participants discussed, on an informal basis, several different aspects of the economy, the business and the expectations regarding these issues. They also discussed their personal experiences and forecasts. As this group consists of employees from Chinese, British, Australian and German/Dutch companies, the meeting was quite a lively one. One of the participants was the head of the export division of the aforementioned Chinese manufacturer of special building materials. So, in the next few days, business went on as usual; so far, so good . . .

However, a few months after this meeting, the employee from the German/Dutch company, specialized in developing new business for clients in the international marketplace, was contacted by the Chinese employee by email, with the following request: 'Do you think it is feasible to participate and sponsor a conference in the Pacific, on the basis of which we can expand our brand more internationally? See the attached announcement. Please feel free to comment.'

After first wondering why they had contacted just him, he wrote back and asked the reason. They explained as follows: 'Because you were showing a quite open approach during the discussion at the dinner last year, after the conference when we met. You had an interesting view on the market with a strong sense of commitment, on the basis of which we think you could help us in growing the business.'

Following this, as a business development specialist he knew of course about the dos and don'ts in working with new clients: first, a clear goal should be defined as a consistent strategy; second, not least, what resources would be available for them to finance this project? So in short, was this a serious request for an interesting (consultancy) project and/or opportunity to establish a new joint business?

The practice: international case studies 139

Nevertheless, he knew that this would not be the right way to communicate his thoughts. So, while keeping alert and checking their seriousness positively, he replied in a polite and professional way that their thoughts about this new conference marketing were really interesting, but might not be as efficient as expected; for example, how to organize the follow-up? Was the foreseen audience the right client-base? How to deliver if interested prospects showed serious reponse? And so on. After sending this quite polite reaction, he took some time to get a response, but it was affirmative: yes, the marketing budget could indeed be used in better ways than this. His suggestions were taken seriously, and they would keep considering next steps.

This was indeed also a very polite and supportive reponse, but with no real issues to connect to for working towards a new business; so what to do now? He had checked for himself the size and profile of the Chinese company, and these were interesting enough to add to the client-base for new business development. As a specialized business developer in international businesses, he therefore decided to act positively and proactively, just as the Chinese employee had done. He made a proposal to them, consisting of the following issues:

- to work together on the basis of a joint deal, thus trying to open up new markets together;
- and on a long-term basis within a strategic (international) partnership;
- but with the support of the top management.

Although the business developer was appointed to work on a number of different international markets, the export markets considered for this case study were specifically Hong Kong and Macau. These two markets, very close to mainland China and even part of the mainland, are still interesting because of their differences from mainland China regarding policy and social system. Both regions are considered special economic zones, based on their previous foreign-owned structure (Hong Kong was British and Macau was Portuguese). Therefore the added backgrounds for this case study are as follows:

- expansion in the Hong Kong and Macau market;
- based on establishing partnerships with (existing) distribution channels.

Using existing distribution channels (which already distribute other non-competing products) is a quick way to open new markets for new products. That was considered also to be the case for the Hong Kong and Macau markets. However, although the Hong Kong and Macau market is quite small compared with mainland China, it still has differences regarding specific characteristics, for example in the area of technical standards and norms. This also influences the possibilities for exporting to these markets. Figure 4.17 shows several ongoing building activities within the dynamic city of

140 *The practice: international case studies*

Hong Kong, illustrating Hong Kong also as a city, considered to be one of the 'economic gateways' to mainland China.

Figure 4.18 depicts a typical feature of the Hong Kong and Macau construction market is depicted: the use of bamboo scaffolding. Although this type of scaffolding equipment is often used in other regions of Asia also, it shows an interesting contrast between, on the one hand, the 'high-tech' designs of the local office projects and, on the other hand, the use of such a 'low-tech' type of equipment; however, it may indeed raise the consideration that using a very suitable type of equipment such as bamboo scaffolding can still also be considered as quite 'high-tech': a very suitable and easy to handle type of equipment, perfectly fit for purpose, and in those regions a quite easily available, fast-growing plant, thus obviously also a quite sustainable solution!

Although in general the total deal structure described in this case study was not too complex, the experiences during the 'construction process' to get into the total deal were still quite interesting. This case study describes and analyses these experiences in more detail.

Figure 4.17 Ongoing building activities in Hong Kong, illustrating Hong Kong also as a city, considered to be one of the 'economic gateways' to mainland China.

Figure 4.18 The use of bamboo scaffolding is typical of the Hong Kong/Macau and Asian construction markets.

Stakeholders involved

Several stakeholders were involved during the project. The following stakeholders are mentioned, including their roles:

- **Client**: a large Chinese company, producing special building materials. Its production factories were located in several locations in mainland China. Because it delivered to international markets only through project-based contacts, its strategy was increasingly focusing on export activities by using a more structured distribution approach.
- **Business developer**: a German/Dutch company, specialized in developing new business for clients in the international marketplace.
- **Distributor**: as part of a shortlist, preselected by the business developer, this company was a locally based Hong Kong party, with experience in distribution of other non-competing building products.

Differences in behaviour of the stakeholders involved

One of the most important specifically requested issues was creating a kind of framework for the foreseen deal. Although working together on the basis of a joint deal and within a strategic long-term (international) partnership was also important, the support of the top management was essential, especially because it is always better to have business deals supported by the responsible top management of the company (thus preventing any possible future misunderstandings regarding such a deal, etc.). However, looking more closely at this situation it was indeed absolutely necessary, because in general in China the control mechanisms and management structures (within the collectivistic Chinese business culture; Hofstede, 1991) are generally said to be based on a 'simple structure' (Mintzberg, 1983), a type of business control that requires top management to confirm all decisions taken. This in contrast to, for example, the UK ('adhocracy'), Germany ('professional bureaucracy'), France ('complete bureaucracy') and the USA ('division structure'). It is understandable that this Chinese characteristic is related to the background of the communist era, during the early days of which chairman Mao Zedong stated that individualism was not allowed; it was all about collectivism, with personal supervision and a strategic top organization.

In present-day China this culture of collectivism sits alongside a strong degree of individualism. However, looking in more detail at business-cultural research issues, the Chinese situation is still quite interesting: it is a society that especially balances long-term and short-term goals. This is also known as 'Confucian dynamism', referring to the ancient times of the famous Chinese master Confucius. This is seen as represented in behaviour, for example as the continuous balancing between 'truth' and 'virtue' (Hofstede and Bond, 1988). Other ancient Chinese masters are often still referred to in modern business literature; another example is the famous Chinese general Sun Tzu's work about war strategies. His strategies are also often used for lessons on basic negotiation strategies, often seen as comparable to 'being in war' (DENMA, 2009).

In this specific project the total 'construction process' of preparing the planned distribution deal went as follows.

The business developer had made a preselection list of suitable parties able and willing to establish the planned distribution structure for the Chinese building products. In a quite logical approach, he had made a list in order of preference. First on that list was a possible distribution candidate recommended to him by good business friends within his strong Asian network. So he checked the suggested distributor and found out about its good reputation and positive suitability for this foreseen new dsitribution market. After this preselection, he started up the next steps:

1 Getting the approval of the Chinese headquarters to continue in this way, which also meant getting the exclusivity basis for this potential distributor reconfirmed regarding the Hong Kong and Macau market. As is understandable, the potential distributor asked explicitly about reconfirming the exclusivity basis. So at this point there was no problem, as it was going as agreed between the business developer and the Chinese producer.

2 Parallel to this, the business developer had arranged a non-disclosure agreement regarding this total approach with the potential distributor, so that the distributor would not disclose any information about the strategy, business networks of the producer and so on.
3 Exchange of information regarding detailed material specifications, reference data about products and former projects in mainland China and the like.
4 Organizing a joint visit to visit the production factories, offices and so on, and to have a closer look and face-to-face negotiations on how the potential distributor could really get into a total distribution agreement with the producer.

Steps 1, 2 and 3 were completed just before the summer holidays. The correspondence and telephone calls between the business developer and the producer were going smoothly and satisfactorily, as also between the business developer and the potential distributor. So there seemed to be nothing to block the fourth step (i.e. organizing the visit and starting the further negotiations with the potential distributor 'on site' in China).

Shortly before the summer holidays started (which was for all parties more or less the same period), the visit proposals for matching the agendas were exchanged, resulting in some possible joint dates. The business developer had already scheduled the planned visa application and so on, so there were no signals about any disturbance regarding this organized schedule.

Immediately after the summer holidays the business developer contacted the producer to reconfirm the agreed travelling schedule, which was agreed again by the management representatives of the export department. So he also reconfirmed it with the intended new distributor, who also started to organize the travelling arrangements. However, a few days after this, the business developer received a polite but very clear email from his business partner, the Chinese producer, explaining to him more or less as follows:

- Yes, the visit to the production facilities and so on was initially agreed.
- Yes, they looked forward to a good collaboration with the intended new distributor.
- However, exclusivity could no longer be guaranteed.
- Based purely on that fact, they realized themselves that, in this changed situation, the intended distributor might well no longer be interested in organizing the Hong Kong and Macau markets.
- This might make the foreseen visit quite useless in this case.
- However, just in case, if requested, the planned visit could still be organized.

Nevertheless:

- As the management representatives of the export department, they felt of course very sorry about this unexpected development.

144 *The practice: international case studies*

- However, they could not do anything about it, because their CEO had in the meantime (during the summer holidays) agreed on an alternative solution, offered by another possible distributor which their CEO had obviously already known long before.

The business developer felt of course very uncomfortable with this suddenly changing situation. Through the correspondence and telephone calls they had immediately after this unsatisfactory message, he confirmed that the export department representatives of his Chinese business partner felt uncomfortable too about this changed situation. They therefore continued to explain and apologize to him that they had tried but could not change the situation any more. As a way of solution, they offered him better commercial conditions if he brought in other markets for them. So it became clear to him that in fact they did not want to harm the situation and damage the existing strong relationship (i.e. *guanxi*) with him. That was a positive signal despite their negative move regarding the Hong Kong and Macau markets.

However, the business developer felt that his real problem now was the relationship with the preselected distributor candidate. That relationship was based on existing relationships with other parties in Asia, who had recommended him to this experienced and suitable distributor. How to bring the bad news to them? Combined with this, the positive situation was that he could still offer the possibility of organizing the distribution in other Asian countries too, as a possible alternative to the Hong Kong and Macau markets.

He decided just to call and email them, in that order. As expected, the foreseen distributor felt very uncomfortable, too. So there was a lot to explain for all parties involved, but he managed to do so by also contacting in parallel his other Asian business friends, explaining to them too what had happened. The business developer proposed that they consider the alternative opportunities of getting the distribution in other Asian countries; however, after thinking a while they decided not to go forward with that. As a result, they therefore decided to cancel the planned trip to mainland China, and would not enter into any other distribution markets for the Chinese producer. The business developer accepted this, and showed his respect to them, which in turn also resulted in an ongoing good relationship, despite this unsuccesful business opportunity.

So the business developer and his Chinese business partner were now free again to look jointly for new distribution markets and possible distributors. However, in the meantime the Hong Kong and Macau markets were opened by the other distributor, leaving the business developer with empty hands, but nevertheless with a better commercial deal for him for the other markets to develop further.

Examining the case as a whole, the professional parties involved in the project represented in general the following behaviour:

- **Client**: The client acted according to the strict manner of polite correspondence and agreed structure of deal making. However, although

the management of the export department had made the arrangements, their CEO suddenly changed their policy for export markets, thus harming the position of the business developer within their joint partnership. Nevertheless, in return, they offered him a better deal for other future markets, trying to keep the relationship (i.e. *guanxi*) as good as possible.
- **Business developer**: Being experienced in Asian business traditions, he recognized the changed positions of the stakeholders involved, but he also understood that it was necessary to give some 'space' for this movement, thus creating a stronger 'credit' for his future joint businesses there.
- **Distributor**: As a party also used to working within a business culture based on trust at the one hand and opportunism on the other hand, he quite quickly understood the situation. However, keeping to his initial strategy, he did not change his strategy regarding which markets to be active in. The result was that he decided to stay out of this possible business deal.

Gradually it became clear that the behaviour of specifically two main parties was the main reason for disturbing the construction process:

- **Client**: Despite the fact that the correspondence and discussion were going through the management of the export department, the central CEO was still the decision-maker in these. Although the business developer had arranged contracts for other earlier markets too, signed by the same CEO, there obviously still existed a kind of overall company strategy which might not fully match the specific export strategy.

    Comment: Business culture in present-day China is characterized by collectivism combined with a strong degree of individualism. However, looking in more detail at business-cultural research issues, the society seems especially to balance between long-term and short-term goals, which is also known as 'Confucian Dynamism' (Hofstede and Bond, 1988). From that point of view, it is not strange that the CEO chose to follow the short-term opportunity offered by a quick deal with another distributor within their own existing network, instead of a long-term solution through the contacts established by the business developer. From the business developer's point of view, the timing might then also be one of the issues, because of the delay caused to the initial visiting schedule by the intervening holiday period. However, to restore trust, the management of the export department confirmed to him that their CEO's action was indeed obviously just caused by the influence of some direct external contacts of the CEO, probably combined with following a kind of short-term opportunity: in their eyes, the business developer had not done anything wrong at all; he therefore should not feel sorry, and, as a way of restoring mutual confidence, he would be compensated extra by them in following deals for them within other regions.
- **Business developer**: Based on his earlier experiences, his way of trusting

his Chinese client and business partner seemed to have become somewhat overbalanced. This, combined with continually requesting reconfirmation of existing agreements and information exchanged, might give the impression of feeling uncomfortable and/or showing a kind of distrust towards the business partner. So he went on in preparing the planned deal structure, obviously without being alert enough to possible 'weak signals' sent by his Chinese business partner, which might be signs that parallel influencing issues were going on.

Comment: As the relationship between the business developer and his Chinese client and business partner had already lasted for a relatively long time, it would in general not be a logical thing to refresh it. However, because it was not possible to be continually in each other's physical locations, in this case it was a relationship at a physical distance; and this seems to be increasingly the case in nearly all business relationships, especially because of the increased use of modern media with all its adavantages but undoubtedly also disadvantages (Tijhuis, 2002). However, to keep such international business relationships 'alive', he felt afterwards that a continual 'refreshing' of the business relationship would have been better during the preparation of the planned deal structure. So, during the preparation of the subsequent deals with them about other new markets, he based the way of working on an increasing exchange of information, shared decision making and reconfirmations, not as a sign of distrust, but as a means of keeping the mutual trust and growing the day-to-day commitment for working on joint deals. Although this approach often seems to be quite a balancing act (Tijhuis, 2004), it obviously is still an effective way of working for the parties involved.

Based on the 3C-Model™, one can represent the above situation as 'types of behaviour', being:

- positive (i.e. speeding up the running process);
- negative (i.e. slowing down the running process).

In this case the negotiations did not lead to a deal for establishing a distribution structure for the Hong Kong and Macau markets, because of the suddenly changing strategy by the top management of the Chinese client; they at once followed other existing contacts, which obviously emerged suddenly during the ongoing negotiations between the business developer, the intended distribution party and the client. From that viewpoint the process was indeed negatively disturbed by this unforeseen behaviour of the client. However, from another viewpoint, it also meant a promise of better future deals for the business developer in other new markets to come. So instead of a complaint by the business developer, the solution for this emerging conflict was based on an immediate promise of a serious compensation by the client for the unsatisfactory situation, thus behaving mutually within the context of maintaining the existing long-term relationships. This approach was a satisfactory solution for maintaining progress then, and avoiding a real conflict situation. In Figure 4.19 this is represented schematically in the three phases of the 3C-model™.

Although the use of a compensating promise was felt as a satisfactory solution for the existing situation, such promises are of course taken seriously only if the relationship between the parties involved is already established and firm, and felt to be on a mutual basis. This was the case in this situation. For new relationships this way of behaviour will be experienced as a quite adventurous way of doing business. Nevertheless, anyone in these situations should consider the fact that Chinese and Asian business cultures in general are quite cautious about losing their good reputation but are flexible about working according to certain procedures, whereas Western European business cultures are quite flexible with their good reputations but cautious about working apart from procedures; this means, generally speaking, a difficult built-in conflict. However, if the parties involved are being somewhat more flexible about working according to procedures but more strict on keeping their good reputation, this may stimulate the creation of a sustainable long-term business relationship with a win–win approach for all the parties involved.

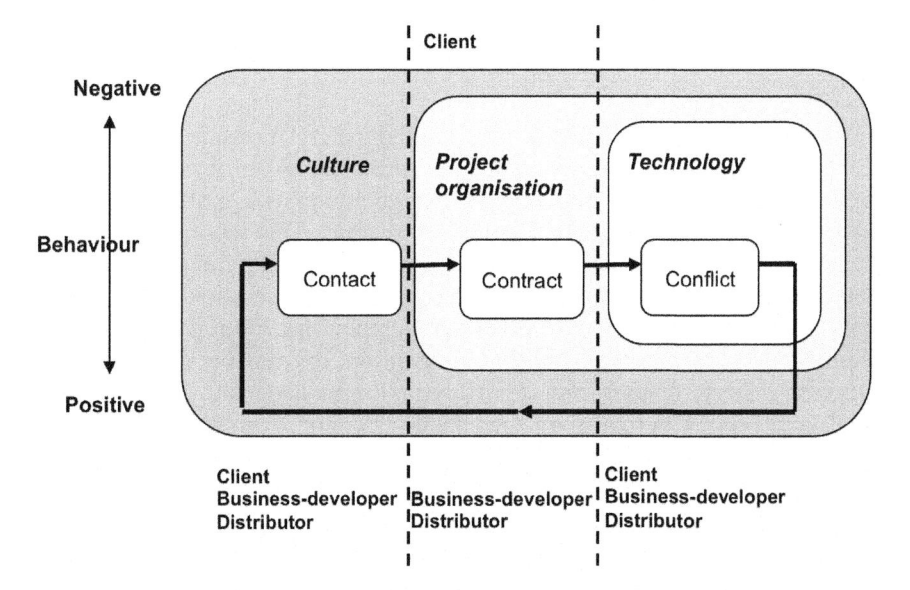

Figure 4.19 Types of behaviour (positive and negative) of the parties involved in the three phases of the 3C-Model™.

148 *The practice: international case studies*

Conclusions and recommendations: lessons learned

Contact (culture)

LESSON 1

When establishing a business relationship within a business environment where strong personal and values are active, this means especially a strong need for manoeuvring between several sensible issues, for example formal and informal relationships, family issues, political issues. Parallel to this the existence of a (large) physical distance between the parties involved plays an important role, which cannot easily be solved by just the use of social media or whatever. In such environments the risk for changing patterns within existing relationships is quite high. This means that parties involved should be constantly balancing and refreshing the starting points and shared basic values, to see if mutual interests are still satisfactory according to their common goals. Therefore as lesson 1 is defined:

> Creating business relationships in itself is not particularly difficult. The most important issue is the maintaining of such business relationships, especially by a continuous balancing of the shared interests and goals. This is especially the case when business partners work at a long physical distance, as is often the case in, for example, international business environments.

Contract (project organization)

LESSON 2

In this situation, the existence of a contractual relationship between the Chinese client and the German/Dutch business developer obviously did not prevent the client from following an emerging parallel opportunity for the new Hong Kong and Macau distribution market. This was quite a typical action, because he decided to do this despite still feeling himself bound by the contract, as he showed by offering a good compensation for the business developer with respect to other coming joint distribution markets. The business developer knew, from that moment on, the way he had to balance between short-term goals and opportunities on the one hand, and long-term goals and reputations on the other hand. Therefore lesson 2 is defined as follows:

> Existing contracts between parties do not mean that the arrangements made in such contracts are always followed clearly. Especially in business cultures with a focus on short-term goals, an opportunistic approach is often used to follow alternative approaches outside the

contracts. However, if parties can balance this method by a continually focusing on long-term goals and keeping the value of reputations, this still can lead to a satisfactory situation for all parties involved. Nevertheless, this will often need a business culture in which the having and keeping of a good reputation plays an important role.

Conflict (technology)

LESSON 3

This case study was in fact a 'perfect' example of a situation, representing several ingredients for getting into a real conflict: offending against an existing contract, unilateral action by one of the business partners, despite the existing joint agreed team approach, and so on. In fact this behaviour is more or less reflected in the following saying: 'In decision making it is often better to ask for forgiveness afterwards than to ask for approval beforehand'.

However, the business developer, as offended party in this, behaved wisely because examining the situation he immediately asked how he would be compensated, instead of only focusing on trying to get them to change their decision. This was wise especially because the influencing factors were important: the alternative distributor was obviously a good friend of the top management; that meant that offending such personal relationships would not help in solving the existing situation in the short term. Leaving the possibility for the client to compensate the business developer through another distribution market in fact saved the client's reputation, and satisfied the business developer in the long term, too. Therefore the focus on creating solutions with a long-term focus instead of focusing on discussing who is guilty was also experienced here as a way to avoid falling into conflicts. Lesson 3 is therefore defined as follows:

> For the prevention of conflicts there is a need for a joint focus of the parties involved, especially by trying to search for solutions instead of just discussing who is guilty. However, this needs the parties involved to give space for such solutions and possible alternatives, together with the focus on a realistic time period in which such solutions can be realized, and thus be able to satisfy the involved parties.

Resumé

Keeping trust in certain contractual relationships is often a difficult issue, because it implies that everything which cannot be written down or included within the meaning of a contract is covered by trust. Although often practised, it is no guarantee of preventing breaches in the contractual arrangements, because in business situations people are acting not only formally but also informally, often following their emerging short-term opportunities.

Nevertheless, keeping the value of the existing good reputations in the front end of all the decision-making processes, including the possibility of giving space for alternative solutions in case of difficulties regarding existing contracts, can really help to prevent falling into conflicts. However, this will need parties whose business cultures match regarding these issues, or at least which understand each other.

Resumé

In this fourth chapter several experiences within the described case studies were analysed with the help of the presented 3C-Model™. It showed several differences of behaviour between the stakeholders involved, which can be seen by focusing on 'snap-shots' of critical moments negatively influencing the progress of the construction processes. This follows the recommendations of Schein (1985) and Sanders (1995), who both more or less recommend focusing on critical incidents in the practitioner's situation if one wants to learn what the influence is of business cultures and how they represent human behaviour in daily (construction) processes. For this book a few of many available examples have been selected, described within the case studies, on the basis of access to in-depth experiences, although one can distinguish several other examples in which the progress of such ongoing construction processes is influenced negatively, because of the daily complex situations between several stakeholders involved within construction processes.

These case studies are therefore especially meant to give an illustration of the daily complex situations within the construction industry regarding the existence and influence of business cultures and differences between them, not only as examples for practitioners, academics or students, but for all who want to learn more about the influence of these business cultures in daily construction practice, improving the awareness for this interesting part of the daily business and the human behaviour of the stakeholders involved.

5 Lessons learned

Introduction

The selection of the six case studies described in this book has led to an overview of lessons, which are wrapped up in the first section of this chapter. These lessons are listed and combined per analysed theme of the 3C-Model™, and described as lessons for:

- sustaining contacts;
- improving contracts;
- preventing conflicts.

The resulting overview is used to describe a summary of these lessons, which are altogether meant as recommendations for practitioners, academics and students, so that they can use them for their daily and/or future activities.

Lessons for sustaining contacts

An overview

The following seven lessons have been defined:

a A serious attention to the culture (behaviour, reputation, etc.) of possible parties involved when establishing the project team should therefore be recognized within this selection process. The 'match' with its project environment should be taken seriously. Trying to understand team members' 'hidden agendas' as well as their 'official agendas' is of great importance.
b Good communication at the front end of the process, involving all the stakeholders, would possibly have improved the acceptance of the project, at least if the architect and developer had accepted their influence at an early stage.
c Trust is good, but control is better; even when parties know each other quite well, there is still a serious risk of changing behaviour owing to

the influence of a changing situation within the construction process. Nevertheless, too much control can be counter-productive because of the risk of frustrating the parties involved. Therefore, there is especially a need for a good balance between trust and control.
d Competition within a team is good, to keep every team member's approach focused on the client's goals. However, before selecting and/or exchanging team members, one should carefully investigate the match between these party's attitudes, thus keeping a grip on the team (group) dynamics.
e Even if parties do have their national background in common, their behaviour within a multi-national group or consortium is not automatically an equal one; this is because every individual party will work according to its own agenda, which may be different from that of the other parties involved, regardless their own or others' nationality. Therefore, the matching of team members should be done especially with the focus on matching the goals and agendas.
f Working together with different specialist parties in a team needs a strong sense or awareness of the weak signals related to the behaviour of the different parties involved. In particular, knowing each party's professional and/or private backgrounds and drive can reduce the risk of unpredicted behaviour during the total construction process.
g Creating business relationships is not in itself particularly difficult. The most important issue is the maintaining of such business relationships, especially by a continuous balancing of the shared interests and goals. This is especially the case when business partners work at a long physical distance, as is often the case in, for example, international business environments.

Summary

Altogether, the overview of the lessons for sustaining contacts can be characterized within the following three key lessons:

- *Communication from the very beginning* within project organizations and accompanying processes is essential for creating a sound basis for a succesful project. Being aware of the influence of the business-cultural background of the parties involved seriously affects such processes.
- However, one should not make the mistake of thinking that such business cultures create a static pattern of behaviour by the parties involved. The difficulty is that they often result in a *dynamic changing behaviour* of the parties involved during the process, which introduces a serious need to develop a sound sense – or a kind of antenna – for weak signals.
- Alltogether this works the best by focusing on *long-term business relationships*, so that one can really learn to know the parties involved. Nevertheless, one should always be aware that several parties are not at all interested in work within long-term business relationships.

The resulting three keywords are:

Communication – Dynamic – Long-term.

Lessons for improving contracts

An overview

The following seven lessons have been defined:

h Keeping a clear separation of the several roles within projects, divided among the parties involved, is always possible. However, when deciding this in the contract, each of the separate parties should have roles, being related to at most one side of the 'three sides of the table' (i.e. client/developer-related, or neutral-based or contractor/subcontractor-related).

i Choosing and deciding about team members for a (building) contract should be not only on a legal or financial basis, but also based on the experiences and 'culture' of the parties involved.

j The decision for choosing the type of contract should be based not only on the characteristics of the project itself, but also and especially on the balancing of the involved participants' characteristics. Parallel to this, the decision on accepting the contract should be made not only on the basis of the project characteristics, but also and especially on the basis of the 'suitability' for one's own organization.

k Contracts are nowadays a necessary tool for getting projects realized. However, one should be very sensitive to the weak signals of the parties involved, during the process of negotiations before, during and after the signing of the contract. This especially because construction is still a people business, which means that project results cannot just be guaranteed by contracts.

l The procedures used by clients for tendering for specific contracts should be fully clear to parties before they are invited to join such tender procedures. This to avoid unclarity because of uncontrolled negotiation processes, which will not lead to a satisfactory result for all the parties involved, but will instead lead to a waste of time.

m Although trusting the parties involved within their business environment can be recognized in general as a positive signal for the specific situation between these parties, it should never lead to unclear situations regarding their mutual agreement: Is there really commitment? Did parties sign for what they expected it to be? And so on. Therefore, in business environments in general, dealing on the basis of mutual trust is good, but it should be so good that it is just logical to add the signatures to this mutual trust as well.

n Existing contracts between parties do not mean that the arrangements made in such contracts are always followed clearly. Especially in business cultures with a focus on short-term goals, an opportunistic approach is often used to follow alternative approaches outside the contracts. However, if parties can balance this method by a continually focusing on long-term goals and keeping the value of reputations, this still can lead to a satisfactory situation for all parties involved. Nevertheless, this will often need a business culture in which the having and keeping of a good reputation plays an important role.

Summary

Altogether, the overview of the lessons for improving contracts can be characterized in the following three key lessons:

- Within contracts *one should avoid more than one role per party involved*. Such contracts should be chosen not just on the basis of the project characteristics, but also and especially on the basis of the characteristics of the parties involved: if the project characteristics match the characteristics of the relevant organization.
- Contracts in themselves are not a guarantee of smooth processes and pleasant business relationships. They have merely to do with the (level of) trust, and especially with *keeping the capability to maintain the good reputations of the parties involved*.
- The influence of business culture within contracts has especially to do with *balancing between short-term opportunities and long-term relationships*. Nevertheless, if one is able to keep this balance properly, it can still result in a satisfactory situation for all the parties involved, despite (gradually) moving away from the initial goals of the contract.

The resulting three keywords are:

Roles – Reputations – Relationships.

Lessons for preventing conflicts

An overview

The following six lessons have been derived:

o Besides the project team members, an open and structured communication (a 'dialogue') during the total construction process is an effective instrument for taking away certain fears of stakeholders (with active and less active roles) in projects. This certainly can reduce the risk of disturbing projects in an early stage, although it could require extra attention and energy on the part of the team members. But that's worth it.
p Although one is more comfortable with a personal way of handling conflicts, one should be prepared to accept that the juridical way will often become the final means; this when working with professional as

well as private parties. This may also be an extra reason to be very selective during acceptance or rejection of projects and their accompanying contracts and business partners.

q During negotiatons for getting problems solved, one should especially keep the focus on those issues which connect parties, instead of only focusing on those issues which are just separating parties. This will increase the chance of finding solutions by routes and/or issues of mutual interest between the parties involved.

r For solving conflicts it is important to focus on the present situation and future results of actual decisions to be taken. However, these decisions are often perceived by other parties involved based on their previous experiences during comparable situations. Knowing their behaviour during such comparable situations and/or projects within their background can help to keep the conflict situation under control, and in working towards a feasible solution. However, one should always be aware of (unexpected) actual circumstances influencing the behaviour of the parties involved, making the prediction of such behaviour quite difficult and complex.

s Within several projects the existing business relationships lead to conflicts. However, such conflicts are better viewed as an opportunity to obtain a better understanding of the underlying problems and clarify the situation; and this especially if seen as a critical incident, in which conflicts can even be of use for improving one's capabilities for handling these difficult situations, thus leading to better prevention of such conflicts in future. Nevertheless, especially in business situations one should always be able to separate the business conflict from the personal relationship.

t For the prevention of conflicts there is a need for a joint focus of the parties involved, especially by trying to search for solutions instead of just discussing who is guilty. However, this needs the parties involved to give space for such solutions and possible alternatives, together with the focus on a realistic time period in which such solutions can be realized, and thus be able to satisfy the involved parties.

Summary

Altogether, the overview of the lessons for preventing conflicts can be characterized in the following three lessons:

- For better avoidance of conflicts, *serious communication is needed at an early stage*, although this may need a lot of energy and attention to be paid for a well-established dialogue within processes and projects.
- In the process of solving conflicts, it is extremely important to *focus on those issues which connect the parties involved*, instead of just focusing on those issues which are separating the parties involved. However, to reach a feasible result and convince all the parties involved, such attitude

should be combined with a serious time-schedule for setting next steps.
- Always keep in mind that a conflict is not just a negative issue; it also gives the parties involved the opportunity to learn, because during such moments one is behaving 'as one really is', generally speaking. This also implies the *need to know the background – business culture – and earlier experiences of the parties involved*, to be able to make a sound forecast of their behaviour in the actual or emerging conflict.

The resulting three key terms are:

Early stage – Connections – Early experiences

Resumé

Viewing the summary of the lessons from the described case studies and the additional analyses in relation to the influence of business cultures on construction industry, the need to pay serious attention to this issue seems to be quite evident, not least because the globalizing construction industry is increasingly influenced by a growing number of international projects and parties, all searching for a piece of the cake, thus resulting in an increase in competition. To be able to distinguish oneself within this globalizing market, it is increasingly necessary to know not only one's own chararacteristics, but especially also the characteristics of the competitors in the market. Business cultures, differences between them and their influences on daily processes and projects within the construction industry are important examples of such characteristics. If one is aware of that importance for one's daily business, it can help practitioners to improve their competiveness in the globalizing marketplace. In parallel to this, it can challenge academics and students to learn more about the ins and outs of business cultures and their influence on (construction) business.

6 A future vision for culture in international construction

Introduction

In the scope of this book, the focus in the described case studies was merely on critical incidents, as suggested by Schein (1985) and Sanders (1995) in their research on business cultures and organizations. This approach has proven to be a useful one, as negative moments are often useful examples to learn from, especially based on the assumption that human behaviour during such moments in general cannot hide itself, because human nature obviously has a kind of built-in mechanism to defend oneself against negative situations, thus implying that the real behaviour will be exposed especially during such negative moments.

To give further food for thought on the future role of business culture in the construction industry, the following three issues will be highlighted in this chapter:

- developments in the construction industry;
- developments in behaviour during critical incidents;
- business culture's role in the construction industry.

These three issues are described and analysed in more detail and with a brief description in the following paragraphs of this chapter. Although the authors analyse current developments in the construction industry that are influencing human behaviour, for example the rise of corporate-level stand-alone or internet-based collaborative software tools, internet-based search engines and, more recently, the rise of social media networks, especially supported by internet and mobile communication structures, the authors are still aware that they cannot predict the future. However, the analyses are based on realistic assumptions, so that they can paint a realistic vision to give practitioners, academics and students basic viewpoints to develop further.

Developments in the construction industry

In the present-day construction industry several influences can be distinguished. Think, for example, about several ups and downs within the general economy, heavily influencing also the market of the construction industry, thus proving that construction is acting as a kind of 'thermometer' for the economy, being very sensitive to the influences of the economy in general, especially in the housing sector, as in general for directly consumer-related business (Tempelmans-Plat, 1984). It often plays a role in boom times, leading to, for example, real estate 'bubbles', as was the case in the former East Germany during the approximately five years immediately after the collapse of the Berlin Wall in 1989 (*Der Spiegel*, 1994). However, in combination with such strongly growing real estate markets, one can also often see the strong decline as the 'bursting bubble' that (logically?) follows (*Der Spiegel*, 1995).

Parallel to this, the construction industry itself had and has its own problems as well, caused by, for example, 'typical' behaviour in this sector. See for example the recent need for several revaluing programmes for this sector since the mid-1990s in the United Kingdom, focusing on 'team-building' approaches to improve the often conflicting client–contractor relationships and so on (Latham, 1994; Egan, 1998). Other examples during the early 2000s are governmental investigations regarding collusion discovered between several (groups of) contractors in tenders for public infrastructure projects in the Netherlands; since then governments have strongly focused on adapting and improving the regulations, to eliminate such practices in this sector (Vos, de Wit, Duivesteijn, van der Staaij, Smulders, van Beek, Pe and ten Hoopen, 2002).

Generally speaking, the construction industry and its products (for example real estate and infrastructure projects) are obviously still an attractive sector for investors not only because of the large amounts of capital needed to realize these type of projects, but obviously also because of its quite complicated structure of many small, middle-sized and large organizations, working together for their (groups of) clients, and often in a quite untransparent way.

The exchange of information and communication structures within this industry therefore plays an important role for understanding and managing these type of businesses, thus helping to improve the practices in construction industry; and presumably not only in this industry, but also in other types of industries. This not only on a national scale, but also on an international and multi-national scale (Bomers, 1976), where the influence of different business cultures may arise even more strongly within an international scope, as was, for example, also seen in the previously described case studies.

Although the authors are aware of the fact that a further list of possible influences might become endless, for example regarding sustainability issues and the need for poverty reduction in developing countries, or the increasing need for flexible housing concepts with integrated intelligent ambient technology for the ageing society (see, for example, Du Plessis, 2002; Vita Valley and TNO, 2010), they selected the following shortlist, mainly to keep within the scope of this book by focusing on 'drivers' for influencing personal communication and behaviour within construction industry:

A future vision for culture in international construction 159

1 growing use of information and communication technology (ICT);
2 globalization of the construction industry.

These two development issues are analysed in more detail in the following sections.

Growing use of information and communication technology (ICT)

Modernizing information and communication technology (ICT) is increasingly influencing the several parties involved in construction. In general, the use of ICT tools is even seen as one of the serious ingredients for overcoming crisis situations within the construction industry, as part of (international) recovery strategies; this is the outcome of a large international investigation among practitioners and academics by Ruddock and Ruddock (2010). Nevertheless, for a real recovery of the sector, Tijhuis (2009) points at the fact that recovery strategies should not just be based on a hit-and-run approach, as often seen in the construction industry, for example its focus on solutions for day-to-day problems. Within this perspective, recovery strategies should especially be fuelled by a clear focus on the mid- to long-term approach, so that they can really lead to a sustainable improvement of unsatisfactory (crisis) situations.

Going into more detail regarding ICT tools, their introduction within working environments in general leads to a reduction in the need for human labour capacity, which is generally perceived as a negative influence. See, for example, the influence of subcontracting practices on working relationships and employment, where there seems to be an increase in temporary contracts instead of contracts of employment, which is also influenced by the gradual opening of international labour markets (Raiden *et al.*, 2004).

However, the authors still see the use of ICT as a positive influence and absolutely necessary for maintaining the international competitiveness of construction parties, especially because it reduces, for example, the manual transfer of information between parties, thus reducing and even avoiding the risk of unacceptable changes and/or failures in information (for example reduction of failure costs). ICT-driven tools such as BIM (Building Information Modelling) that combine 3D drawings with integrated databases of technical/commercial information are gradually superseding the traditional combinations of (3D) drawings with additional separate technical/commercial information and/or documents, thus creating, in essence, an 'nD interorganizational project environment'. Such a way of working has been gradually developed, tested and used on several projects in the international construction industry. However, despite all the described advantages for construction parties, such approaches as the use of interorganizational ICT still encounter some barriers in the business environment which have to be overcome; parallel to this, it is also obvious that drivers have to be

160 A future vision for culture in international construction

introduced for stimulating the use of ICT in a broader perspective within the construction industry.

In relationship to these issues, Adriaanse and colleagues (2010) investigated the mechanisms influencing the external motivation to use ICT in construction projects; these are the results of observations within an existing project environment where parties used the so-called BIM approach. They identified two main influencing mechanisms:

- availability of contractual arrangements about ICT use;
- presence of requesting actor.

Their findings of influencing mechanisms as results of the test projects are represented in Figure 6.1.

Within the scope of this book it is interesting to see that the 'requesting actor' (as representative of the human factor), notwithstanding any enforcing contractual arrangements, still plays in an important role in deciding whether or not to use ICT

Although CAD systems in general work on the basis of 3D approaches, the additional information in the digital designs often consist of, for example, cost information, material specifications, time-schedules, purchasing and contractual obligations. In general, the 3D approach is expanding within the BIM approach towards a more or less 'nD approach'. Several developments

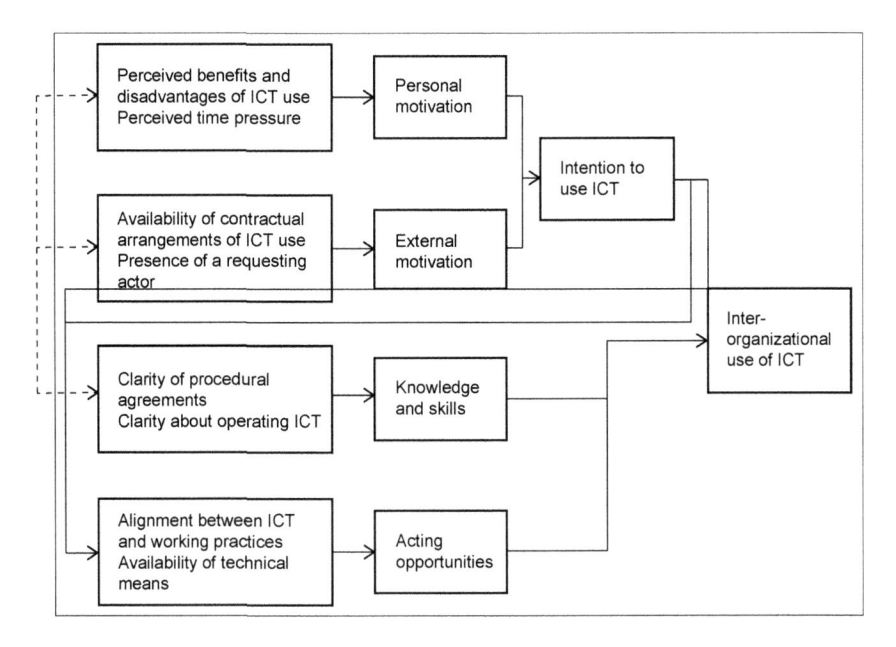

Figure 6.1 Mechanisms influencing external motivation to use ICT in the construction industry (Adriaanse *et al*., 2010).

A *future vision for culture in international construction* 161

in this field are ongoing within an international scope, for example since the British Construct I.T. initiative dating from the early 2000s (Construct I.T., 2002). Additional initiatives are focusing on developing process models as optimized structures of decision making, also functioning as process 'layers' or 'formats' for the development of software tools. The so-called 'process protocol' is such an initiative, offering a basic structure for shaping a clear information exchange and communication structure within construction processes (Cooper and Aouad, 2002).

Within this context, Van Rees describes the computer as a 'partner in construction'. In his research on several practices involving ICT tools and their use within today's dynamic construction industry, he emphasizes the important role of ICT (van Rees, 2007). At the same time, he also states that for successful innovation in the construction industry nowadays it is of paramount importance to eliminate the human role within the information 'translation' process, viewing humans as a common bottleneck within information-related processes.

In relationship to such bottlenecks, it is still interesting to read that in early analyst reports about the so-called 'e-commerce' developments, it was stated that:

> the real story will lie not in the matching of buyers and sellers on the internet but in automated fulfillment solutions, strategic collaboration and optimization. This will require robust, reliable, secure and highly flexible solutions that can deal with complex real-time transactions and rapidly changing business processes.
> (Westhead, Mortenson, Moore and Williams Rice, 2000: 1)

From such a viewpoint it is more or less the ability to deal with rapid changes in business which constitutes the barrier and/or driver for the use of ICT.

Nevertheless, despite the above-mentioned barriers and drivers for the use of ICT, the developments of such ICT tools is still going forward, with or without their use in the construction industry. An interesting assumption might be then that (near!) future generations will increasingly do their work according to their 'gaming' behaviour. Why shouldn't they? After all, they are used to playing their games consoles, featuring sophisticated games, incorporating pretty advanced examples of sophisticated simulation software and platforms. Companies which are more or less acting on the cross-links of such developments are, for example, some of the existing leisure companies, developing their facilities increasingly in an integrated way by using sophisticated design and simulation software tools. A well-known company in this field is Disney®, which has named its integrated design approach as a way of 'imagineering' (Schwegler, 2009). Several other parties in construction are also using multi-dimensional ('3D' or even so-called 'nD') approaches for their design and engineering software tools, including project datastructures (for example BIM), increasingly integrated with, for example, web-based

162 A future vision for culture in international construction

process management tools. Several actual projects can be mentioned as positive representatives of such practical (3D-engineered) project results, also and especially in difficult project situations. The new terminal building at Letisko International Airport in Bratislava, Slovakia, completed in 2010, is such a positive example of the efficient use of information technology tools during the design and engineering work preparation and realization process. Construction activities on existing and operational airports are often quite difficult, because of the continuous effects of ongoing (air) traffic on and around the building site. This underpins the need of thoroughly rationalized design and engineering principles, not only to deliver future capacity and flexibility during the foreseen use of the building, but also supported and integrated with an efficient and safe construction logistics process during the realization period. Figures 6.2, 6.3 and 6.4 show the existing building, the building site during realization and the completed new terminal building at Letisko International Airport in Bratislava (PIO KMP and NTGroep, 2009/2010).

These examples show also the growing influence of, for example, integrating movements, focusing on bringing together several data streams such as (mood- and/or decision-influencing) social media networks with technical or design-oriented tools; from that point of view the integration of 'everything'

Figure 6.2 The existing building at Letisko International Airport in Bratislava, Slovakia (PIO KMP and NTGroep, 2009/2010).

Figure 6.3 The building site for the new terminal building at Letisko International Airport in Bratislava, Slovakia, designed, engineered and realized by the efficient use of information technology tools (PIO KMP and NTGroep, 2009/2010).

leads in fact to a kind of 'mesh collaboration', stimulating the creation of new business value, and leads also to other ways of decision making by changing clients' behaviour within such changing collaborative networks (Mulholland and Earle, 2008). Above all, the (near!) future generations are not only the managers and employees within the (near) future, but also the governments – and clients! Therefore a client-focused approach should look not at the barriers but merely at the opportunities which such new technologies can bring.

Not only the changing of the clients' behaviour, but also the way in which the collaboration between companies is evaluated, is characterized by a more or less increased 'collaborative approach', parallel with the fact that competition in general is getting stronger within (construction) markets. This collaborative approach is also supported by a push and/or pull approach by the increasing availability of sophisticated software and hardware tools. These developments are represented by, for example, the earlier published vision of one of the early ERP (enterprise–resource–planning) software pioneers, Baan, whose developments within the so-called 'beyond-ERP' approach have created, for example, a platform for a collaborative software architecture (Baan, 2005). His innovative vision and approach

Figure 6.4 The completed new terminal building at Letisko International Airport in Bratislava, Slovakia (PIO KMP and NTGroep, 2009/2010).

resulted in a software platform, called Cordys™, which has become one of the first challenging examples of sophisticated corporate-level stand-alone and internet-based collaboration tools: it is able to function in fact as an 'orchestrator' between existing software packages within and between companies, making it possible to integrate the data-flows within existing separate software packages, and manage these data-flows by, for example, defined key performance indicators ('KPI') through 'dashboard' functions. This approach reduces the need to replace such existing software packages by just focusing on the collaborative aspects and sophisticated interfaces, such as input/output management tools.

Such collaborative software tools, based on a kind of service-oriented architecture (SOA) and functioning on a stand-alone corporate level as well as on a web-based structure, are increasingly stimulating the way of processing data within and between companies through, for example, the so-called 'cloud computing' approach: in fact processing a company's data-flow not only using software installed on internal company computers, but through software available online. These types of approaches are on the one hand reducing the need for local computing power, but on the other hand leading to a stronger demand for internet capacity and security, and a need for greater reliability. However, also from the viewpoint of such approaches, the internet more or less is, and seems to be increasingly, 'the' backbone for clients, companies and other parties involved, functioning as 'the' exchange platform and management tool for their businesses and communication; and this not only regarding managing facts and figures, but also and increasingly regarding managing opinion and behaviour.

The increasing popularity of internet-based search engines and especially the social media networks, supported by internet and mobile communication structures, exemplifies how popular 'tools' can influence clients' opinion and behaviour; and it is not only clients that are influenced by such developments, but also other people involved directly (employees, bankers, etc.) or indirectly (potential clients, neighbours, business analysts, etc.) in the present-day construction business. In the following paragraphs these developments will be analysed in more detail.

Globalization of the construction industry

Borders between countries are increasingly diminishing, if not geographically, in a political and economical sense. The rise of developing regions such as Brazil, Russia, India and China (the so-called BRIC countries) adds an enormous potential of consumer spending to the world markets. These regions formerly were separated and/or divided by different political systems and economic traditions, but nowadays their governments understand the influence of the growing cross-border communication and thus trading activities. An increasing outsourcing of production, fuelled by the still existing differences between labour and production costs in the different regions of the world, is one of the results of these developments. In parallel, it is also stimulated by increasing transportation capacities and modalities, created and supported by, for example, the strong developing (container) shipping activities around the world.

However, the globalizing market also stimulates the development of increased competition, because of the diminishing borders, and supported especially also by such innovations as improved ICT tools. This acts as a driver for a true 'reformation' of businesses today, as Fingar (2006) described, and related to this he also highlighted Abraham Lincoln's statement:

> The dogmas of the quiet past are inadequate to the stormy present. The occasion is piled high with difficulty and we must rise with the occasion. As our case is new we must think anew and act anew.
>
> (Fingar, 2006: 15)

This statement may also reflect on the present-day construction industry. A small example in a construction-related environment (toys) may be the world-famous Lego® blocks. Young children and grown-ups all around the world play and build with this standardized building system. From that viewpoint it was quite logical that such types of blocks would experience fierce competition from competitors with their imitation brands, trying to tap into the success of the building system. However, because of the costly juridical implications of breaching patents and trademarks, resulting in battling against a large worldwide company, those competitors with their imitations were not really successful. That was a lucky and prosperous situation for the market leader, Lego®. However, since the recent ending of official protection by patents and trademarks of the design of the blocks ended, a totally new competition situation has arisen in that market, with a new situation regarding the position of the market-leader: how to keep the top position, if everyone is allowed to copy the blocks (Fd.nl, 2010)? A challenging new time for the (toy) building system may arise, indeed needing a new way of thinking and acting for the market leader as well as for its competitors.

Especially after the further commercialization of the World Wide Web, accompanied by the bursting of a huge ICT bubble in 2000, for example, it is nowadays nearly unthinkable that construction industry businesses would function without such web-enabled technologies. See also the issues described above. However, looking more closely at early analysts' reports describing the upcoming e-commerce platforms for the construction industry in those days, it is interesting to note that several of those 'early adopters' were not able to achieve significant growth in the construction market, at least from a European viewpoint (Stockdale, Campbell, Naslin and Wong 2000).

This low growth level, and even decline, of those early adopters and also the rise of new e-commerce parties can probably be related to the view of Stockdale and colleagues during those days; they saw room for only two or three pan-European independent marketplaces as the industry develops its own e-commerce strategy. However, another more important reason for this might be also in their other described view, stated as follows: 'we do not expect lower pricing to result from the development of B2B, due to the local/regional nature of the industry and the adverse value to weight ratio of many products' (Stockdale *et al.*, 2000). Altogether these are interesting developments, in which geographical borders are rapidly diminishing in businesses, thus stimulating international expansion of the (construction) industry. However, apart from differences between (business) cultures and the local circumstances, the quite heavy construction materials used, besides the differences between standards and regulations across different countries, still make it difficult to act on a global scale. Nevertheless, the globalization of information processing is rapidly growing, although there too the differences between (accounting) regulations and codes [for example, IFRS (International Financial Reporting Standards) and US GAAP (United States Generally Accepted Accounting Principles)] make it still difficult to 'export' existing business models and processes; the increase in the use of cross-border cloud computing (Mulholland *et al.*, 2010) may even be slowed down more or less by such aspects. So harmonization of regulations on an international scale is an important issue for making globalization a successful business approach or not.

However, regardless of the possibilities offered by technology, harmonizing regulations and so on, there will always remain the important factor in (construction) business:

A future vision for culture in international construction 167

Human beings . . . !

Without them, no *client*, no *funding*, no *design*, no *construction*, no *users*, no . . . ? One might probably add 'no fun' but also 'no problems': two important aspects often lacking in the male-dominated construction industry. From that viewpoint, Figure 6.5 shows a relatively unusual sight in the construction industry: a *laughing* and *female* construction professional on a construction site.

Problems obviously often occur during construction processes, as was illustrated in the case studies in this book. Nevertheless, the authors see in the near future not only negative impacts on behaviour as illustrated, but also positive impacts. The following paragraphs briefly describe their opinion and outlook on that future.

Developments in behaviour during critical incidents

As was seen in the case studies, the parties involved in construction projects quite often come into conflict with each other. Such conflicts often arise from issues related to, for example:

- calculated pricing compared with costs;
- construction time and planning/scheduling;
- delivered and expected quality levels;
- injuries and lack of safety on sites.

Figure 6.5 A relatively unusual sight in the construction industry: a laughing and female construction professional on a construction site.

168 *A future vision for culture in international construction*

Generally speaking, the way people behave differs throughout different situations, because the individual and group influences, combined with past experiences, will influence the resulting behaviour of the individual. Individual behaviour – whether part of a group or not – and the surrounding environment is still the important issue to focus on if one wants to avoid conflicts. Verma points at the way individuals are able and willing to strive for a compromise in such possible conflict situations by balancing the desire to satisfy oneself with the desire to satisfy others. This so called 'Thomas–Kilmann' model is represented schematically in Figure 6.6 (Thomas, 1976; Verma, 1996).

In the present-day construction industry one would expect that there are many occasions to balance the influences of all parties involved, for example because of the several moments of decision between and during every phase of the total construction process. Thus, from initiative and design until delivery, operation and re-use of a construction project, there are several opportunities to prevent or reduce the eventual emergence of critical incidents between the parties involved. However, taking an overview of the several publications and news items regarding such critical incidents between, for example, professional parties in the construction industry and their professional as well as private clients, one can conclude that the nature of this industry and/or the nature of the parties involved somehow does not

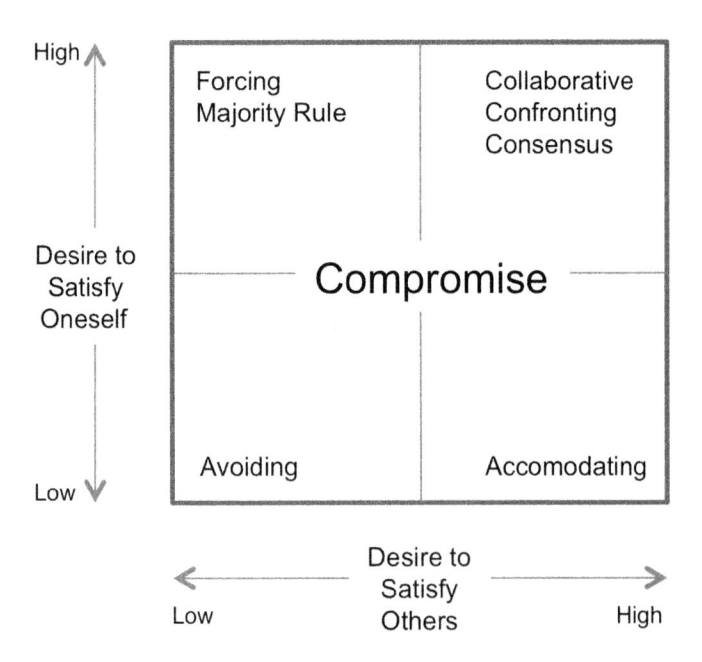

Figure 6.6 The 'Thomas–Kilmann' model, showing the way a compromise in a possible conflict situation is reached or not (Thomas, 1976; Verma, 1996).

A future vision for culture in international construction 169

stimulate the prevention and/or reduction of such critical incidents. At least, that seems to be the overall impression.

Analysing the critical incidents in construction industry within the scope of this book, the business culture (i.e. nature) of the parties involved still seems to be one of the most important reasons for such emerging critical incidents. Therefore the need to match the business cultures between the parties involved before definitively deciding about the possible start of a joint construction project is of strong importance.

Matching such business cultures according to the described 3C-Model™ (Tijhuis, 1996, 2001a) will therefore help to handle construction processes: not only the organizational consequences (contracts) or technical consequences (conflicts), but also and especially to begin with the cultural consequences (contacts). That the serious attention to and alertness for specific behaviour and its weak signals in the (construction) industry is still of great importance is also shown by several recent publications of construction- and real estate-related conflicts, also related to construction projects. Some of them even led to scandals, caused by the criminal behaviour of individuals involved; for example, in the Middle Ages, during the construction of St Peter's Basilica in Rome (Scotti, 2006), or more recently during the development of several prestigious office projects in Amsterdam (van der Boon and Van der Marel, 2009). A better attention to several weak signals regarding the behaviour of individuals would probably have had a preventative effect in such cases. Apart from specific criminal behaviour of certain individuals as above, the research of Deenstra, for example, showed that, on the other hand, the regular professional behaviour of, for example, project managers cannot automatically be linked to the level of overall success of the projects which they are managing (Deenstra, 2001). This altogether makes the following assumption quite realistic:

> Business culture (in construction) has the characteristics of a kind of Trojan Horse: it's there, but to a large extent one cannot exactly predict the outcomes of business culture's influences on daily (construction) business processes.

Nevertheless, staying alert and being aware of the influence of business culture on daily business processes will help to prevent and to solve problems with it on every organizational level, from the production or construction site to the top management.

This suggested awareness of the influence of business culture is needed more explicitly, because the people involved are getting more and more interconnected, as shown, for example, by the increasing influence of internet-based search engines and social media networks. These developments altogether are really influencing the behaviour of (groups of) people, by connecting them, influencing their opinion and/or probably even harmonizing their behaviour. In fact it indicates the following important influencing factor in the construction industry (and certainly not only in this type of industry) for developments of (patterns of) human behaviour within present and near future business environments:

> Social media networks and other internet-based tools mobilize 'hidden' forces within (groups of) people, especially between the people directly involved in the (construction) business (clients, employees, bankers etc.), but also and especially between the people indirectly involved in the (construction) business (potential clients, neighbours, business analysts etc.).

Although the rise of the strong influencing power of internet-based search engines since *c.* 2000, for example Google™, Yahoo™ and their early Chinese competitor Baidu™, has been quite impressive, the founding and growth of several present-day social media networks such as Facebook™, Twitter™, LinkedIn™ and Hyves™ since *c.* 2004 has been even more impressive, at least when comparing their speed of (national/international) growth and increasing number of (global) users. As an illustration, several authors have published their analyses of the founding and rise of some of those companies, trying to illustrate in one way or another the actual influence and characteristics of those companies (e.g. Vise and Malseed, 2005; Kirkpatrick, 2010). One could debate the truth of such analyses (which are often unofficial and unauthorized accounts of those company's historical backgrounds); however, there is no doubt about the enormous power encapsulated within the worldwide community of those networks' users. Because of, or despite, their strengths and weaknesses, these search engines and social media networks have an increasingly influential role, thus in one way or another also leading to changing (patterns) of human behaviour. As a result, these behavioural changes are also influencing seriously the companies themselves in the (international) business environment, including their business models with all their strengths and weaknesses. The following two examples illustrate some of the emerging opportunities as well as threats, for example in the situation of a contractor company and its client:

- If a contractor company has one unsatisfied client, then that is a serious problem, which needs to be solved properly. However, if that client is an active user of social media networks and/or the internet, he or she might feel the need to publish his or her opinion about the company on the social media network; and in such a way numerous people worldwide can easily read about the bad experiences with that contractor company, thus probably harming the company's reputation seriously. So, if the contractor company does not act properly and solve the problem promptly, the (indirect) damage, as a result of the harm to the company's reputation, may be even much higher than the costs of repairing the delivered unsatisfactory project.

A future vision for culture in international construction 171

- When clients get increasingly interconnected, they can use their increasing purchasing power to start bargaining with several (groups of) contractors, suppliers and so on to buy large amounts of materials, standard types of projects and so on. This type of empowerment of clients is actually used in, for example, the Groupon™ website, using certain time-related purchasing periods (for example per day) as an opportunity for joint online bargaining with suppliers of certain products. It also gives the opportunity for companies and suppliers to generate large amounts of sales of (standardized) products, probably with lower margins, but with higher sales volumes. This is probably merely a trend towards a business model based on a 'daily supermarket' approach.

In general, the market (experiences etc.) and market price (costs, margins etc.) are becoming increasingly transparent to the extent that existing business models (generally based on the availability of 'hidden' information) are under threat; on the other hand, this will give opportunities to new business models (generally based on open sources of quality and pricing information).

Nevertheless, despite the positive or negative influence of *true* information, distributed via social media networks and/or the internet, there is also the large risk of spreading positive or negative *false* information. Because all types of information have a certain positive or negative impact on people, both companies and individuals should always be aware of what kind of information about them is circulating on the social media networks and/or internet, especially because removing information still seems to be quite difficult, even years after the first digital upload of information.

Specific governmental regulations might help to improve the existing situation. However, as these are global issues, global harmonization may also be required. So this is probably an interesting (global) challenge, although it seems more and more that the technological developments are going extremely fast, whereas the accompanying regulations etc. simply cannot be adapted properly within the same time-frame. This continuously leads to a kind of grey area in which several technological developments still can expand, but in which certain public or private interests still bear the risk of being harmed in one way or another by these developed technologies.

Figure 6.7 illustrates these developments schematically, representing social media networks as a tool for interconnecting people, thus stimulating influencing opinions and/or harmonizing human behaviour of (groups of) people.

Nevertheless, one should always remember that, as an individual or a company, one stays part of the social network, despite choosing to use or not use social media networks or other internet-based tools. However, the latent risk with such tools is still in the fact that they can also easily be abused, by publishing secret information, for example, probably harming certain public or private interests, or false information. The case of Wikileaks™ illustrates

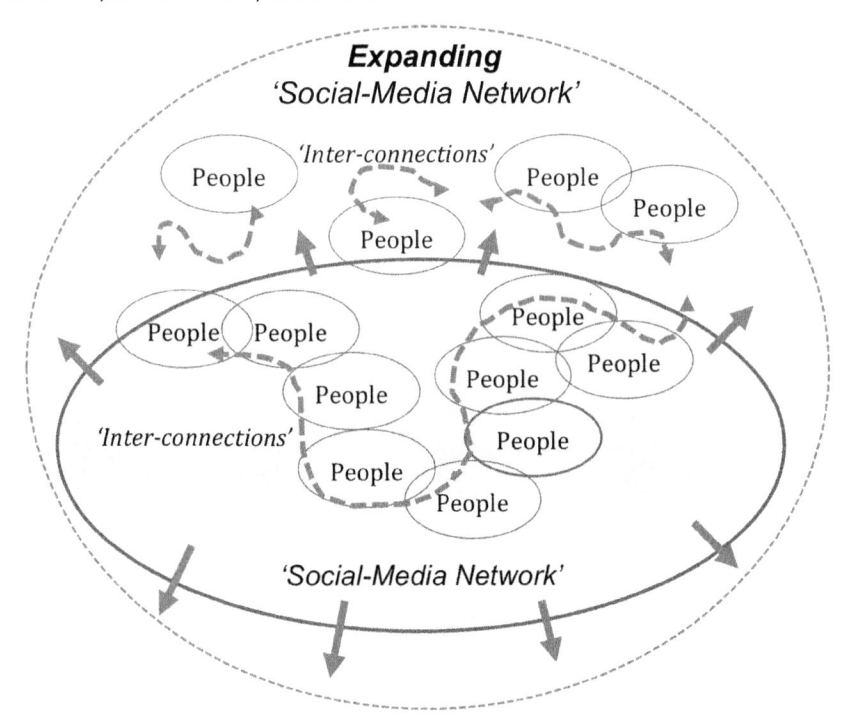

Figure 6.7 Schematic representation of social media networks as a tool for interconnecting people, thus stimulating influencing opinions and/or harmonizing human behaviour of (groups of) people.

the effects of publishing secret information on a national and international scale and shows that a proper use of social media and related tools needs an honest and clear way of acting. Wikipedia™ is another example, it not always being clear about what listed information in the encyclopaedic online database is true or false. These examples also indicate that today's society in general – as (groups of) users and/or producers of information and communication – may really need to rethink the rules of the game – probably a challenging task for all.

Business culture's role in the construction industry

In the present-day construction industry the influence of the existing business cultures in different situations is still evident; people's behaviour still plays a significant role in the process of negotiation, deal making, conflicts, mediation and so on. But what will be the role of business cultures in the future? Will they continue to influence construction processes? Will people continue to behave according to the 'built-in' characteristics of their business cultures? These and other such questions cannot be answered precisely. However, the authors do expect an increasing awareness by the people involved regarding the existing influence of business cultures in construction industry. From that perspective, business culture is gradually growing from a 'black box' towards a more transparent issue with step-by-step discoveries of patterns of behaviour, accompanying circumstances and ways of running construction processes. This is an encouraging development, because in the early days when the authors were both involved in the CIB Task Group TG23, founded in 2000, the overall assumptions and attitude in construction industry were that business culture-related issues were merely a 'soft' item, and not to be regarded as at all important in the 'hard' construction processes. As one of the signals of the TG23's success it was seen that in 2005 it was officially upgraded towards an official international working commission: CIB W112 'Culture in Construction', in which both the authors were involved as joint coordinators. The publication of a report discussing the theme of 'Culture in Construction: Part of the Deal?' highlighted an interesting discussion between construction professionals and sociologists, more or less setting the trend towards a paradigm-shift for the business culture theme: from being a non-influencing item towards a strong influencing item in the present-day construction industry (Tijhuis, 2001b).

Business culture and globalization: harmonization versus differentiation

An interesting accompanying effect is the current developments and effects of globalization: where one would expect that differences between cultures are diminishing because of the gradual diminishing of political borders (for example in the European Union), at the same time one can see that individuals and groups of people somehow are trying to keep their own specific culture alive or even reshape their culture towards a more integrated version of multi-cultural behaviour characteristics. However, it seems that the integration of multi-cultural behaviour towards a unified 'standardized' way of behaviour is not feasible; the present situation is the existence of several groups of people side by side, behaving and working according to their own 'traditional' culture, not only in their private life but also and especially in their professional life. This is to a large extent the situation in the European Union: although, for example, technical standards and norms are increasingly harmonized, the different interpretations of specific rules make it still quite difficult to create a common and equal understanding and handling of specific situations, for example when trying to solve or prevent conflicts.

Therefore, an ongoing research and documentation of examples and case studies of culture's influence in construction industry can help practitioners, academics and students to learn more and to understand better the impact of culture's influences in their daily business, for example by categorizing described case studies and discovering patterns in displayed behaviour. That will be helpful for all the parties involved, because the construction industry is a people business, and it is expected to stay that way.

An interesting parallel can be seen within the medical sector: although the theories and medications for curing diseases are more or less harmonized in an international scope, there are still differences in interpretation on how to use certain procedures and (local) knowledge and experience for the healing process. There are regions in the world that completely rely on the help and effect of (chemical) medicine, but there are also regions in the world that rely purely on a kind of witch-doctor, using traditional beliefs, local (natural) ingredients and so on (Morley and Wallis, 1978). However, the two different situations do not have the exclusive rights on effective curing: in fact both situations can be effective and positive. Nevertheless, to be alert to the influence of those (local) characteristics, and aware of their role as probably an important barrier or driver for a specific curing process, will help to understand the patient, and can even help to prevent diseases.

When looking at the construction industry again, this could be 'translated' to a comparable assumption as follows:

> The construction industry (the patient) needs to be cured by tailor-made treatment (processes), which needs a good knowledge and understanding of the local behaviour and beliefs (business culture). Using this properly, it will help to solve conflicts (diseases) and even to avoid them.

Altogether it leads to a quite amazing situation that, although communication tools such as language, websites, telephones and emails are increasingly harmonized, the individual users still behave according to their own principles and backgrounds. Not only in society in general but also in the construction industry this seems to be largely the case. Therefore the well-known ideas 'think globally, act locally' and 'it's better to know someone in the market than just having market knowledge' seem to remain very important basic understandings for success in international business; not only in the present-day construction industry, but also and especially in the near future. Being aware of this makes it also clear that the risk of misbehaviour of individuals within the construction industry or within other businesses will stay a latent but nevertheless still serious risk for all the parties involved.

In addition to this, the previously described increasing influencing power on (groups of) people by, for example, internet-based search engines and social media will surely play an important role in that near future; it empowers such (groups of) people to act according to their daily need within certain situations, even on a harmonized global way, by influencing their opinion and ways of acting. Altogether this reduces also the possibility of predicting such human behaviour on the basis of a certain behavioural history, that is according to certain business-cultural patterns. This also supports the assumption that the only certainty regarding the near future is that *human behaviour will change*, but it is uncertain what will be the pattern of that expected change. Nevertheless this is an interesting challenge and need for construction industry, for example by continually trying to better match expectations and specifications between parties involved, thus serving the client within a more satisfactory way, presumably leading towards a more profitable business too.

Resumé

As a resumé of this book, the authors conclude that during recent years the influence of business culture on daily construction industry has grown increasingly important not only in international projects, but also in national projects. This despite the fact that, for example in the European situation, the EU countries' political borders are gradually diminishing, whereas parallel with this harmonization the differences between the several groups of people, living in the several geographical areas, seem to be increasingly pronounced. Other developments such as the described ICT influences are connecting the world more and more towards one global marketplace. To learn more and to understand better culture's influence in the construction industry, an ongoing research and documentation of examples and case studies can help practitioners, academics and students in their daily business, especially because the construction industry is expected to stay a people's business, and all those people, whether *directly* or *indirectly* involved, are increasingly getting interconnected by, for example, internet-based search engines and social media networks, with all their strengths and weaknesses, also challenging the construction industry to handle and use the emerging opportunities and threats properly now and in the near future.

It is in fact comparable to developments in medicine and the healthcare industries, where it is well known and accepted that the methods and effectiveness of curing interventions in human healthcare also depend on and differ between the several cultures in which the people (i.e. patients) are living. Projected onto the construction industry, this especially means that a stronger focus on (differences between) business cultures in the construction industry ('patient's background and behaviour') will improve the total business process ('organism') and solve and even help to prevent conflicts ('diseases'). Nevertheless, always be aware that business culture still acts as a kind of Trojan Horse, with largely unpredictable influences on daily business processes, at every level of an organization, from the production or construction site to the top management.

This leads to the following conclusion:

Culture in construction has become increasingly part of the deal!

No one involved in the construction industry can avoid culture's influence, so everyone involved should be better aware of it and use its influence positively.

Notes

1 Introduction

1. There are exceptions to that approach, such as in replicating populations in samples for tests on the safety of medicines.
2. Thanks to Dr J. Gunning for reminding the authors.

References

Adriaanse, A., Voordijk, H., Dewulf, G. (2010) Adoption and use of interorganizational ICT in a construction project, *Journal of Construction Engineering and Management*, **136** (9), 1003–1014.
Agenda 21 (1992) *Earth Summit Agenda 21 The United Nations Programme of Action from Rio*, http://www.un.org/esa/dsd/agenda21/res_agenda21_07.shtml (accessed 15 February 2006).
Akiner, I. (2005) *Satilik Kültürler: Liberal Mimarlar, Muhafazakar Mühendisler*, Istanbul: Karakutu Yayinlari.
Akiner, I., Tijhuis, W. (2007) Work goal orientation of construction professionals in Turkey: comparison of architects and civil engineers, *Journal of Construction Management and Economics*, **25**, 1163–1173.
Alliance Management International Limited (1999) *Creating Strong Alliances*, http://www.amiltd.com/NewCreating (accessed 14 April 2000).
Alvesson, M. (1994) Talking in organizations: managing identity and impressions in an advertising agency, *Organization Studies*, **15**, 535–563.
Amason, A.C., Hochwarter, W.A., Thompson, K.R., Harrison, A.W. (2000) Conflict: an important dimension in successful management teams, *Organizational Dynamics*, **24** (2), 20–35.
Anderson Consulting (1999) *Outlook Magazine*, http://www.auburn.edu/~boultwr/Hill_Alliances.htm (accessed 21 March 2006).
Au, K.Y. (2000) Inter-cultural variation as another construct of international management: a study based on secondary data of 42 countries, *Journal of International Management*, **6** (3), 217–238.
Baan, J. (2005) *De Weg naar Marktleiderschap: Mijn Leven als Ondernemer*, Amsterdam: Pearson Education Publishers.
Baan, J. (2010) *Business Operations Improvement: The New Paradigm in Enterprise IT*, Putten, the Netherlands: Cordys Holding BV.
Bachmann, R. (2001) Trust, power, and control in trans-organizational relations, *Organization Studies*, **22** (2), 337–367.
Baiden, B.K., Price, A.D.F., Dainty, A.R.J. (2006) The extent of team integration within construction projects, *International Journal of Project Management*, **24** (1), 13–23.
Baumol, W.J. (1959) *Business Behaviour, Value and Growth*, New York: Macmillan.
Beamish, P.W. (1996) The design and management of of international joint ventures, In Redding, G. (ed.) *Cross Cultural Management*, Camberley: Edward Elgar Publishing.

References

Beck, C.H. (1998a) *Polski Kodeks Celny: Polish Customs Code*, Warsaw: Wydawnictwo.
Beck, C.H. (1998b) *Polskie Ustawy Gospodarcze: Polnische Witschaftsgesetze*, 3rd edn, Warsaw: Wydawnictwo.
Beck, W., Herig, N. (1997) *VOB für Praktiker: Kommentar zur Verdingungsordnung für Bauleistungen*, Stuttgart: Richard Boorberg Verlag.
Bernoulli, D. (1738/1954) Exposition of a new theory, on the measurement of risk, *Econometrica*, 22 (1), 23–36.
Black, J.S., Gregersen, H.B. (1992) Serving two masters: managing dual allegiance and expatriate employees, *Sloan Management Review*, 33 (4), 61–71.
Blanken, A. (2008) Flexibility against efficiency: an international study on value for money in hospital concessions, PhD thesis, University of Twente, Enschede, the Netherlands.
BNA, ONRI (2005) *The New Rules 2005: Explanatory Notes on the Legal Relationship Client–Architect, Engineer and Consultant DNR 2005 – Standard Form of Basic Contract*, The Hague: Royal Institute of Dutch Architects and Organization of Consulting Engineers.
Boisot, M., Child, J. (1996) From fiefs to clans and network capitalism: explaining China's emerging economic order, *Administrative Science Quarterly*, 41 (4), 600–628.
Bomers, G.B.J. (1976) *Multinational Corporations and Industrial Relations: A Comparative Study of West Germany and the Netherlands*, Assen: Van Gorcum.
Bon, R., Pietroforte, R. (1990) Historical comparison of construction sectors in the United States, Japan, Italy and Finland using input–output tables, *Construction Management and Economics*, 8 (3), 233–247.
van der Boon, V., van der Marel, G. (2009) *De Vastgoedfraude*, Amsterdam: Nieuw Amsterdam Publishers.
Bordieu, P., Wacquant, L. (1992) *An Invitation to Reflexive Sociology*, Chicago, IL: University of Chicago Press.
Boyer, R. (1996) The convergence hypothesis revisited: globalisation but still the century of nations, In Berger, S., Dore, R. (eds) *National Diversity and Global Capitalism*, Ithaca, NY: Cornell University Press.
BREEAM (2009) http://www.breeam.org/ (accessed 11 August 2009).
Bresnen, M., Marshall, N. (2000) Partnering in construction: a critical review of issues, problems and dilemmas, *Construction Management and Economics*, 18 (2), 229–237.
Burns, T., Stalker, G.M. (1961) *The Management of Innovation*, London: Tavistock Publications.
Cameron, K.S., Quinn, R.E. (1999) *Diagnosing and Changing Organizational Culture*, Reading, MA: Addison-Wesley Longman.
Chandler, A. (1962) *Strategy and Structure*, Cambridge, MA: MIT Press.
Chen, C.C., Meindl, J.R., Hunt, R.G. (1997) Testing the effects of vertical and horizontal collectivism: a study of allocation preferences in China, *Journal of Cross-Cultural Psychology*, 28 (1), 44–70.
Chinese Culture Connection (a team of 24 researchers) (1987) Chinese values and the search for culture-free dimensions of culture, *Journal of Cross-Cultural Psychology*, 18 (2), 143–164.
CIOB, CIB (1987) Managing construction worldwide: three volumes, In Lansley, P.R., Harlow, P.A. (eds) *CIB W65 Conference 'The Organisation and Management*

of Construction'; 5th International Symposium, London, 7–10 September, 1987, London: E & FN Spon.
Clegg, S.R. (1992) Contracts cause conflicts, In Fenn, P., Gameson, R. (eds) *Construction Conflict Management and Resolution*, London: E & FN Spon.
Cole, R.J. (1999) Building environmental assessment methods: clarifying intentions, *Building Research and Information*, **27** (4&5), 230–246.
Construct I.T. (2002) *3D to nD – Developing a Vision of nD-Enabled Construction*, Salford: Salford Centre for Research and Information.
Construction Task Force (1998), *Rethinking Construction: The Report of the Construction Task Force to the Deputy Prime Minister, John Prescott, on the Scope for Improving the Quality and Efficiency of UK Construction*, London: HMSO.
Cooper, R., Aouad, G. (2002) *Process Protocol: Key Principles*, Salford: Development coordinated by University of Salford and Loughborough University.
Dainty, A.R.J., Bryman, A., Price, A.D.F., Greasley, K., Soetanto, R., King, N. (2005) Project affinity: the role of emotional attachments in construction projects, *Construction Management and Economics*, **23** (3), 241–244.
Das, T.K., Teng, B.-S. (1999) Managing risks in strategic alliances, *Academy of Management Executive*, **13** (4), 50–62.
De Boer, L. (1998) Operations research in support of purchasing: design of a toolbox for supplier selection, PhD thesis, University of Twente, Enschede, the Netherlands.
Deenstra, T. (2001) Telt het Gedrag van de Bouwprojectmanager of is het niet te Tellen?, MSc thesis, University of Twente, Enschede, the Netherlands.
Denison, D.R. (1997) *Corporate Culture and Organizational Effectiveness*, Ann Arbor, MI: Denison Consulting.
Denison, D.R. (2009) *Organizational Culture and Leadership Surveys, The Denison Model*, http://www.denisonconsulting.com/advantage/researchModel/model.aspx (accessed 7 October 2009).
DENMA (2009) *The Art of War: Translation, Essays and Commentary by the DENMA Translation Group*, Boston, MA: Shambhala Publications.
DETR (1999) *A Better Quality of Life: A Strategy for Sustainable Development in the UK*, London: The Stationery Office.
DETR (2001) *Building a Better Quality of Life: A Strategy for More Sustainable Construction*, London: The Stationery Office.
Deutsch, M. (1973) *The Resolution of Conflict: Constructive and Destructive Processes*, New Haven, CT: Yale University Press.
Dictionary (2009) http://dictionary.reference.com/browse/industry (accessed 24 July 2009).
Dietrich, M. (1994) *Transaction Cost Economics and Beyond*, London: Routledge.
DIN (1988) *Beton- und Stahlbeton-Fertigteile*, Berlin: DIN (Deutsches Institut für Normung).
Du Plessis, C. (2002) *Agenda 21 for Sustainable Construction in Developing Countries: A Discussion Document*, Rotterdam: International Council for Research and Innovation in Building and Construction.
Egan, J. (1998) *Rethinking Construction: The Report of the Construction Industry Task Force*, London: HMSO.
Eldridge, J., Crombie, A. (1974) *A Sociology of Organizations*, London: Allen and Unwin.
Engineering, Construction and Architectural Management (1997) Special Edition on Private Finance Initiative, **4** (3).

Enshassi, A., Burgess, R. (1991) Managerial effectiveness and the style of management in the Middle-East: an empirical analysis, *Construction Management and Economics*, 9 (1), 79–92.

Euwema, M.C., van Emmerik, I.J.H. (2007) Intercultural competences and conglomerated conflict behaviors in intercultural conflicts, *International Journal of Intercultural Relations*, 31 (4), 427–441.

Ezulike, E.I., Perry, J.G., Hawwash, K. (1997) The barriers to entry into the PFI market, *Engineering, Construction and Architectural Management*, 4 (3), 179–194.

Fd.nl (2010) Lego raakt alleenrecht op bouwstenen kwijt, *Het Financieele Dagblad*, 15 September, www.fd.nl

Fellows, R.F. (1995) *1980 JCT Standard Form of Building Contract: A Commentary for Students and Practitioners*, 3rd edn, Basingstoke: Macmillan.

Fellows, R.F. (1996) *The Management of Risk*, Construction Paper No. 65, Englemere: Chartered Institute of Building.

Fellows, R.F. (2006a) Understanding approaches to culture, *Construction Information Quarterly*, 8 (4), 159–166.

Fellows, R.F. (2006b) Culture, In Lowe, D., Leiringer, R. (eds) *Commercial Management of Projects: Defining the Discipline*, Oxford: Blackwell, pp. 40–70.

Fellows, R.F. (2009) Culture in supply chains, In Pryke, S. (ed.) *Construction Supply Chain Management: Concepts and Case Studies*, Chichester: Wiley-Blackwell, pp. 42–72.

Fellows, R., Seymour, D.E. (eds) (2002) *Perspectives on Culture in Construction*, CIB-Publication no. 275, CIB: Rotterdam.

Fellows, R.F., Liu, A.M.M., Storey, C. (2004) Ethics in construction project briefing, *Science and Engineering Ethics*, 10 (2), 289–302.

Fellows, R.F., Liu, A.M.M., Storey, C. (2009) Values, power and performance on construction projects: a preliminary investigation, In *Proceedings 25th Annual Conference, Association of Researchers in Construction Management*, Nottingham, 7–9 September, CD-ROM.

Felstiner, L.F., Abel, R.L., Sarat, A. (1980) The emergence and transformation of disputes: naming, blaming, claiming, *Law and Society Review*, 15 (304), 631–654.

Ferrell, O.C., Weaver, K.M. (1978) Ethical beliefs of marketing managers, *Journal of Marketing*, 42 (3), 69–73.

Fiedler, F.E. (1967) *A Theory of Leadership Effectiveness*, New York: McGraw-Hill.

Fingar, P. (2006) *Extreme Competition: Innovation and the Great 21st Century Business Reformation*, Tampa, FL: Meghan-Kiffer Press.

Fingar, P. (2009) *Dot.Cloud: The 21st Century Business Platform*, Tampa, FL: Meghan-Kiffer Press.

Fisher. R., Ury, W. (1991) *Getting to Yes: Negotiating Agreement without Giving In*, 2nd edn, New York: Penguin.

Friedman, M. (1970) The social responsibility of business is to increase its profits, *New York Times Magazine*, 13 September.

Furnham, A. (1997) *The Psychology of Behaviour at Work: The Individual in the Organization*, Hove: Psychology Press.

Giddens, A. (1984) *The Constitution of Society: Outline of the Theory of Structuration*, Berkeley, CA: University of California Press.

Glaister, K.W., Husan, R., Buckley, P.J. (1998) UK international joint ventures with the Triad: evidence for the 1990s, *British Journal of Management*, 9 (3), 169–180.

References

Gomez, C., Kirkman, B.L., Shapiro, D.L. (2000) The inpact of collectivism and in-group/out-group membership on the generosity of team members, *Academy of Management Journal*, **43** (6), 1097–1100.

González-Benito, J., González-Benito, O. (2005) An analysis of the relationship between environmental motivations and ISO14001 certification, *British Journal of Management*, **16**, 133–148.

Gruneberg, S., Hughes, W. (2006) *Understanding Construction Consortia: Theory, Practice and Opinions*, RICS Research Paper Series no. 6, London: RICS.

Hagen, J.M., Choe, S. (1998) Trust in Japanese interfirm relations: institutional sanctions matter, *Academy of Management Review*, **23** (3), 589–600.

Hall, P. (1981) *Great Planning Disasters*, Harmondsworth: Penguin Books.

Hall, E.T., Hall, M.R. (1990) *Understanding Cultural Differences*, Yarmouth, ME: Cultural Press.

Handy, C.B. (1985) *Understanding Organisations*, 3rd edn, Harmondsworth: Penguin.

Harris, L.C., Ogbonna, E. (2002) The unintended consequences of cultural interventions: a study of unexpected outcomes, *British Journal of Management*, **13** (1), 31–49.

Hatch, M.J. (1993) The dynamics of organisational culture. *Academy of Management Review*, **18** (4), 657–693.

Hellriegel, D., Slocum, J.W., Woodman, R.W. (1998) *Organisational Behaviour*, 8th edn, Cincinnati, OH: South-Western College Publishing.

Higgin, G., Jessop, N. (1963) *Communications in the Building Industry*, London: Tavistock Institute.

Hillebrandt, P.M. (2000) *Economic Theory and the Construction Industry*, 3rd edn, Basingstoke: Palgrave Macmillan.

Hinman, L.M. (1997) *Ethics: A Pluralistic Approach to Moral Theory*, Orlando, FL: Harcourt Brace Jovanovich.

Hofstede, G.H. (1980) *Culture's Consequences: International Differences in Work-Related Values*, Beverley Hills, CA: Sage.

Hofstede, G.H. (1983) The cultural relativity of organizational practices and theories, *Journal of International Business Studies*, **14** (Fall), 75–89.

Hofstede, G. (1988) Cultural dimensions in management and planning, *Asia Pacific Journal of Management*, vol., 88.

Hofstede, G.H. (1989) Organizing for cultural diversity, *European Management Journal*, **7** (4), 390–397.

Hofstede, G. (1991) *Allemaal Andersdenkenden: Omgaan met Cultuurverschillen*, Amsterdam: Uitgeverij Contact.

Hofstede, G.H. (1994a) *Cultures and Organizations: Software of the Mind*, London: HarperCollins.

Hofstede, G.H. (1994b) The business of international business is culture, *International Business Review*, **3** (1), 1–14.

Hofstede, G.H. (2001) *Culture's Consequences: Comparing Values, Behaviors, Institutions, and Organizations across Nations*, 2nd edn, Thousand Oaks, CA: Sage.

Hofstede, G.H. (2009) http://www.geert-hofstede.com/ (accessed 29 May 2009).

Hofstede, G. (2011) http://www.geerthofstede.com/culture/dimensions-of-national-cultures.aspx (accessed 27 April 2011).

Hofstede, G., Bond, M.H. (1988) The Confucius connection: from cultural roots to economic growth, *Organizational Dynamics*, **16** (4), 4–21.
Hofstede, G., Hofstede, G.J. (2005) *Allemaal Andersdenkenden: Omgaan met Cultuurverschillen*, Amsterdam: Olympus.
Housing Grants, Construction and Regeneration Act 1996, London: HMSO.
HSE (Health and Safety Executive) (2009a) http://www.hse.gov.uk/construction/index.htm (accessed 27 May 2009).
HSE (2009b) http://www.hse.gov.uk/statistics/industry/construction/index.htm (accessed 27 May 2009).
HSE (2009c) http://www.hse.gov.uk/statistics/industry/construction/injuries.htm (accessed 27 May 2009).
HSE (2009d) http://www.hse.gov.uk/statistics/industry/construction/ill-health.htm (accessed 27 May 2009).
HSE(2009e) http://www.hse.gov.uk/statistics/industry/construction/days-lost.htm (accessed 27 May 2009).
Hutton, W. (2002) *The World We're In*, London: Abacus.
Hutton, W. (2006) *The Writing on the Wall: China and the West in the 21st Century*, London: Little Brown.
International Construction (2009) July/August, http://www.khl.com/magazines/international-construction/pdf-archive/ (accessed 30 September 2009).
ISO (2009) http://www.iso.org/iso/iso_14000_essentials (accessed 11 August 2009).
Jemison, D.B., Sitkin, S.B. (1986) Corporate acquisitions: a process perspective, *Academy of Management Review*, **11** (1), 145–163.
Kahn R.L., Katz, D. (1953) Leadership practices in relation to production and morale, In Cartwright, D., Zander, A. (eds) *Group Dynamics*, New York: Harper & Row.
Kahneman, D. (2003) A perspective on judgement and choice: mapping bounded rationality, *American Psychologist*, **58** (9), 697–720.
Kahneman, D., Lovallo, D. (1993) Timid choices and bold forecasts: a cognitive perspective on risk taking, *Management Science*, **39** (1), 17–31.
Kahneman, D., Tversky, A. (1979) Prospect theory: an analysis of decisions under risk, *Econometrica*, **47** (2), 263–291.
Katzenbach, J.R., Smith, D.K. (1993) *The Wisdom of Teams: Creating the High-Performance Organization*, Boston, MA: Harvard Business School Press.
Kipling, R. (2009) *A Truthful Song*, http://www.readbookonline.net/readOnLine/8784/ (accessed 28 September 2009).
Kirkpatrick, D. (2010) *The Facebook Effect: The Inside Story of the Company that Is Connecting the World*, New York: Simon & Schuster.
Koehn, D. (1994) *The Ground of Professional Ethics*, London: Routledge.
Kogut, B., Singh, H. (1988) The effect of national culture on the choice of entry mode, *Journal of International Business Studies*, **19** (3), 411–433.
Korczynski, M. (2000) The political economy of trust, *Journal of Management Studies*, **37** (1), 1–21.
Kroeber, A.L., Kluckhohn, C. (1952) Culture: a critical review of concepts and definitions, In *Papers of the Peabody Museum of American Archaeology and Ethnology*, Vol. 47, Cambridge, MA: Harvard University Press.
Laan, A. (2008) Building trust, PhD thesis, University of Twente, Enschede, the Netherlands.
Langford, D., Hughes, W. (2009) *Building a Discipline: The Story of Construction Management*, Reading: ARCOM, Association of Researchers in Construction Management.

Latham, M. (1993) *Trust and Money: Interim Report of the Joint Government/Industry Review of Procurement and Contractual Arrangements in the United Kingdom Construction Industry*, London: HMSO.

Latham, M. (1994) *Constructing the Team: The Final Report of the Government/Industry Review of Procurement Arrangements in the UK Construction Industry* (The Latham Report), London: HMSO.

Lau, E., Rowlinson, S.M. (2009) Interpersonal trust and inter-firm trust in construction projects, *Construction Management and Economics*, 27 (6), 539–554.

Laufer, A., Raviv, E., Stukhart, E. (1992) Incentive programmes in construction projects: the contingency approach, *Project Management Journal*, 23 (2), 23–30.

Lawrence, P.R., Lorsch, J.W. (1967) *Organization and Environment: Managing Differentiation and Integration*, Boston, MA: Division of Research, Graduate School of Business Administration, Harvard University.

Leary, M.R. (1991) *Introduction to Behavioral Research Methods*, Belmont, CA: Wadworth.

Lewicki, R.J., McAllister, D.J., Bies, R.J. (1998) Trust and distrust: new relationships and realities, *Academy of Management Review*, 23 (3), 438–458.

Lipsey, R.G. (1989) *An Introduction to Positive Economics*, 7th edn, London: Weidenfeld and Nicolson

Liu, A.M.M., Fellows, R., Tijhuis, W. (2002) *A Cross-Cultural Study of Power Disposition in Project Procurement*, RGC proposal, project no. HKU 7021/02E, Hong Kong: Hong Kong University.

Lloyd-Jones, T. (2006) *Mind the Gap! Post-Disaster Reconstruction and the Transition from Humanitarian Relief*, London: RICS, Max Lock Centre at the University of Westminster.

Lucas, C. (2005) *The Philosophy of Complexity*, www.calresco.org/lucas/philos.htm (accessed 7 May 2009).

Lucas, C. (2006) *Quantifying Complexity Theory*, www.calresco.org/lucas/quantify.htm (accessed 7 May 2009).

McGregor, D. (1960) *The Human Side of Enterprise*, New York: McGraw-Hill.

McKenna, E. (2000) *Business Psychology and Organisational Behaviour: A Student's Handbook*, 3rd edn, Hove: Psychology Press.

McSweeney, B. (2002) Hofstede's model of national cultural differences and their consequences: a triumph of faith – a failure of analysis, *Human Relations*, 55 (1), 89–118.

Michel, S., Beuret, M. (2009) *China Safari: On the Trail of Beijing's Expansion in Africa*, New York: Nation Books.

Mintzberg, H. (1983) *Structure in Fives: Designing Effective Organizations*, Englewood Cliffs, NJ: Prentice-Hall.

Moran, E.T., Volkwein, J.F. (1992) The cultural approach to the formation of organizational climate, *Human Relations*, 45 (1), 19–48.

Morley, P., Wallis, R. (1978) *Culture and Curing: Anthropological Perspectives on Traditional Medical Beliefs and Practices*, Pittsburgh, PA: University of Pittsburgh Press.

Mulholland, A., Earle, N. (2008) *Mesh Collaboration: Creating New Business Value in the Network of Everything*, New York: Evolved Media Network LLC; Evolved Technologist Press.

Mulholland, A., Pyke, J., Fingar, P. (2010) *Enterprise Cloud Computing: A Strategy Guide for Business and Technology Leaders*, Tampa, FL: Meghan-Kiffer Press.

Murdoch, J., Hughes, W. (2000) *Construction Contracts: Law and Management*, 3rd edn, London: Spon Press.
Nahapiet, J., Ghoshal, S. (1997) Social capital, intellectual capital and the creation of value in firms, *Academy of Management Best Paper Proceedings*, 35–39.
Negandhi, A.R., Reimann, B.C. (1972) A contingency theory of organization re-examined in the context of a developing country, *Academy of Management Journal*, **15**, 137–146.
Ngowi, A.B., Pienaar, E., Talukhaba, A., Mbachu, J. (2005) The globalisation of the construction industry: a review, *Building and Environment*, **40** (1), 135–141.
Nicolini, D. (2002) In search of 'project chemistry', *Construction Management and Economics*, **20** (2), 167–177.
Odijk, S.W. (2010) Analysis of production of cooled ready-mixed concrete production in an international setting, MSc thesis, University of Twente; Enschede, the Netherlands.
OED (2009) *Oxford English Dictionary* http://dictionary.oed.com/cgi/entry/50259124?query_type=word&queryword=trust&first=1&max_to_show=10&sort_type=alpha&result_place=2&search_id=Psep-Olr6Uv-2754&hilite=50259124 (accessed 26 November 2009).
Office for National Statistics (2009a) http://www.statistics.gov.uk/downloads/theme_commerce/CSA_2008_final.pdf (accessed 16 March 2009).
Office for National Statistics (2009b) http://www.statistics.gov.uk/downloads/theme_economy/BB08.pdf (accessed 16 March 2009).
Office of Government Commerce (2006) *Framework Agreements: OGC Guidance of Framework Agreements in the New Procurement Regulations*, London: OGC.
Office for National Statistics (2003) *UK Standard Industrial Classification of Economic Activities 2003*, London: The Stationery Office.
Ohmae, K. (1990) *The Borderless World*, London: Collins.
Oliver, C. (1997) Sustainable competitive advantage: combining institutional and resourced-based views, *Strategic Management Journal*, **18** (9), 697–713.
Ouchi, W.G. (1982) *Theory Z*, New York: Avon Books.
Pangakar, N., Klein, S. (2001) The impacts of alliance purpose and partner similarity on alliance governance, *British Journal of Management*, **12** (4), 341–353.
PIO KMP, NTGroep (2009/2010) Design-Engineering of Letisko International Airport: Bratislava, project documents, PIO KMP, Trenčín (Slovakia) and NTGroep, Rijssen (the Netherlands).
Pondy, L.R. (1967) Organizational conflict: concepts and models, *Administrative Science Quarterly*, **12** (2), 296–320.
Pritchard, J. (1997) Acting professionally: something that business organisations and professionals both desire?, In Davis, P.W.F. (ed.) *Current Issues in Business Ethics*, London: Routledge, pp. 87–96.
Raiden, A.B., Pye, M., Cullinane, J. (2004) The nature of the employment relationship in the UK construction industry: a multi-tier contracting agreement or a contract of employment?, paper presented to 'Industrial Relations in the Construction Industry', Research Seminar, ESRC/EPSRC, Glamorgan Business Centre, University of Glamorgan, Pontypridd, Wales, United Kingdom.
Ray, S., Chakrabarti, A.K. (2006) Strategic change of firms in response to economic liberalization in an emerging market economy, *International Journal of Strategic Change Management*, **1** (1/2), 20–39.

van Rees, R. (2007) New instruments for dynamic building-construction: computer as partner in construction, PhD thesis, Delft University of Technology, Delft, the Netherlands.

Ren L. (2004) Management of technical innovation in Chinese state-owned enterprises: case studies from a stakeholder perspective, PhD thesis, University of Twente, Enschede, the Netherlands.

Republic of Lebanon (1998) *Progress Report*, Council for Development and Reconstruction, January, Republic of Lebanon.

RIBA (1996a) *SFA Guide: A Guide to the Standard Form of Agreement for the Appointment of an Architect*, London: Royal Institute of British Architects.

RIBA (1996b) *Engaging an Architect: Guidance for Clients on Fees*, London: Royal Institute of British Architects.

van Riemsdijk, M.J. (1994) Actie of Dialoog: Over de betrekkingen tussen maatschappij en onderneming, PhD thesis, University of Twente, School of Business, Public Administration and Technology, Enschede.

Ring, P.S., van de Ven, A.H. (1994) Developmental processes of cooperative inter-organizational relationships, *Academy of Management Review*, 19, 90–118.

Robbins, S.P. (1974) *Managing Organisational Conflict*, New York: Prentice-Hall.

Rooke, J., Seymour, D.E., Fellows, R.F. (2003) The claims culture: a taxonomy of attitudes in the industry, *Construction Management and Economics*, 21 (2), 167–174.

Rooke, J., Seymour, D.E., Fellows, R.F. (2004) Planning for claims: an ethnography of industry culture, *Construction Management and Economics*, 22 (6), 655–662.

Rosenthal, R. and Rosnow, R.L. (1991) *Essentials of Behavioral Research: Methods and Data Analysis*, 2nd edn, Boston, MA: McGraw-Hill.

Ruddock, L., Ruddock, S. (2010) *Emerging from the Global Economic Crisis: Delivering Recovery through a Sustainable Construction Industry*, CIB Publication No. 333, May, Rotterdam.

Samuelson, P.A., Nordhaus, W.D. (2001) *Economics*, 17th edn (international edition), New York: McGraw-Hill.

Sanders, G.J.E.M. (1995) Being 'a third-culture man', *Cross Cultural Management: An International Journal*, 2 (1), 5–7.

Schein, E.H. (1984) Coming to an awareness of organisational culture, *Sloan Management Review*, 25 (1), 3–16.

Schein, E.H. (1985) *Organizational Culture and Leadership: A Dynamic View*, San Francisco: Jossey-Bass.

Schein, E.H. (1990) Organisational culture, *American Psychologist*, 45, 109–119.

Schein, E.H. (2004) *Organizational Culture and Leadership*, 3rd edn, San Francisco: Jossey-Bass.

Schneider, S., DeMeyer, A. (1991) Interpreting and responding to strategic issues: the impact of national culture, *Strategic Management Journal*, 12 (4), 307–320.

Schneider, W.E. (2000) Why good management ideas fail: the neglected power of organizational culture, *Strategy and Leadership*, 28 (1), 24–29.

Schwegler, B. (2009) Opportunities and Challenges of the Use of Interorganisational ICT, keynote speech at workshop held at the University of Twente, 28–30 January, Enschede, the Netherlands.

Scott, W.R. (2001) *Institutions and Organizations*, 2nd edn, Thousand Oaks, CA: Sage.

186 References

Scotti, R.A. (2006) *Basilica: The Splendor and the Scandal: Building St. Peter's*, New York: Viking Adult.

Shakespeare, W. (1964) *King Henry VI Part 2*. Edited by A.S. Cairncross, 3rd edn, London: Methuen.

Shane, S. (1994) The effect of national culture on the choice between licensing and direct foreign investment, *Strategic Management Journal*, 15 (8), 627–642.

Shenkar, O. (2001) Cultural distance revisited: towards a more rigorous conceptualization and measurement of cultural differences, *Journal of International Business Studies*, 32 (3), 519–535.

Sherif, M. (1967) *Group Conflict and Co-operation: Their Social Psychology*, London: Routledge and Kegan Paul.

Sheth, J.N., Parvatiyar, A. (1992) Towards a theory of business alliance formation, *Scandinavian International Business Review*, 1 (3), 71–87.

Simon, H.A. (1956) Rational choice and the structure of the environment, *Psychological Review*, 63 (2), 129–138.

Smith, A. (1789/1970) *The Wealth of Nations*, ed. by Skinner, A., Harmondsworth: Penguin Books.

Der Spiegel (1994) Die Gier frisst das Hirn, In no. 15.

Der Spiegel (1995) Wie von selbst: Der Spekulationsrausch in Berlin ist Vorbei – Viele Buroturme stehen leer, dem Markt droht der Kollaps, In no. 23.

Steenhuis, H.J. (2000) International technology transfer: building theory from a multiple case study in the aircraft industry, PhD thesis, University of Twente, Enschede, the Netherlands.

Steensma, H.K., Marino, L., Weaver, K.M., Dickson, P.H. (2000) The influence of national culture on the formation of technology alliances by entrepreneurial firms, *Academy of Management Journal*, 43 (5), 951–973.

Stockdale, M., Campbell, C., Naslin, S., Wong, A. (2000) *B2B and the Building Industry*, London: UBS Warburg.

Straetement (2002) *Description of Development-Project*, unpublished report, Straetement, Rijssen, NL.

Sugden, J.D. (1975) The place of construction in the economy, In Turin, D.A. (ed.) *Aspects of the Economics of Construction*, London: Godwin, pp. 1–24.

Tagiuri, R., Litwin, G.H. (eds) (1968) *Organizational Climate*, Boston, MA: Graduate School of Business Administration, Harvard University.

Tavistock Institute of Human Relations (1966) *Interdependence and Uncertainty: A Study of the Building Industry*, London: Tavistock Publications.

Taylor, F.W. (1947) *The Principles of Scientific Management*, New York: Norton.

Teece, D.J. (1986) Profiting from technological innovation: implications for integration, collaboration, licensing and public policy, *Research Policy*, 15 (6), 285–305.

Tempelmans-Plat, H. (1984) *Een Bedrijfseconomische Analyse van Bouwen en Wonen: De woondienstenvoorziening beschouwd vanuit een elementenmatrix*, Assen: Van Gorcum Publishers.

Thomas, K.W. (1976) Conflict and conflict management, In Dunnette, M.D. (ed.) *Handbook of Industrial and Organizational Psychology*, Chicago: Rand-McNally, pp. 889–935.

Thomas, K.W. (1992) Conflict and negotiation processes in organizations, In Dunnette, M.D., Hough, L.M. (eds) *Handbook of Industrial and Organisational Psychology*, Palo Alto, CA: Consulting Psychology Press.

Tijhuis, W. (1996) *Bouwers aan de slag of in de slag? Lessen uit internationale samenwerking*, Eindhoven, the Netherlands: Eindhoven University of Technology.

Tijhuis, W. (2001a) Construction industry, e-business and marketing: are the dot. coms pushing aside personal relationships? *International Journal for Construction Marketing*, 3 (1).

Tijhuis, W. (2001b) Culture in construction: part of the deal?, In Tijhuis, W. (ed.) *Proceedings of 2-Day Workshop 22–23 May, 2001 in Enschede, NL*, CIB Report No. 255, Rotterdam: University of Twente.

Tijhuis, W. (2001c) Controlling failure-costs in emerging markets: improving processes without a hit-and-run-approach, In Liu, A.M.M., Fellows, R., Drew, D. (eds) *Proceedings of 'International Conference on Project Cost Management'*, 25–27 May 2001, Department of Standards and Norms, Ministry of Construction, People's Republic of China, Beijing: China Engineering Cost Association, pp. 10, 182–188 (English and Chinese translations).

Tijhuis, W. (2002) Globalizing construction industry and e-business: the value of personal relationships in a dot.com economy, *Journal of Building Construction Management* (formerly *Asia Pacific Journal of Building and Construction Management*), special issue, 7 (1), 50–56.

Tijhuis, W. (2004) Trust and control in procurement: balancing with two dimensions in a collaborative environment, In Kalidindi, S.N., Vargese, K. (eds) *CIB-W92 Symposium 'Project Procurement for Infrastructure Construction', January 2004*, Rotterdam: CIB, pp. 191–200.

Tijhuis, W. (2005) Different approaches in managing complex construction projects – experiences of international project-management, In Carayannis, E.G, Kwak, Y.H., Anbari, F.T. (eds) *The Story of Managing Projects: An Interdisciplinary Approach*, Westport, CT: Praeger, pp. 230–238.

Tijhuis, W. (2009) Crisis or challenge? Thoughts about international construction industry recovery strategies, In Ceric, A., Radujkovic, M. (eds) *CIB-W55&W65 conference 'Construction Facing Worldwide Challenges', 27–30 September 2009, Dubrovnik, Croatia: book of abstracts & CD-ROM*, pp. 57–58.

Tijhuis, W., Fellows, R.F. (2003) Improving construction processes: experiences in the field of contact–contract–conflict, In Bontempi, F. (ed.) *System Based Vision for Strategic and Creative Design: Proceedings of the Second International Conference on Structural and Construction Engineering*, Lisse: Swets & Zeitlinger, pp. 57–64.

Tijhuis, W., Lousberg, L. (1998) TQM and procurement in construction projects: how to handle quality in contractual relationships, In Haupt, T.C., Smith, G., Ebohon, O.J. (eds) *Proceedings of 1st South African International Conference on TQM: 'Total Quality Management in Construction: Towards Zero Defect'*, Cape Town, South Africa: Peninsula Technikon, pp. 25–36.

Transparency International (2009) http://www.transparency.org/news_room/latest_news/press_releases/2008/bpi_2008_en (accessed 7 August 2009).

Triandis, H.C. (1990) Cross-cultural studies of individualism and collectivism, In Berman, J.J. (ed.) *Cross-Cultural Perspectives, Nebraska Symposium on Motivation 1989*, Lincoln: University of Nebraska Press, pp. 41–133.

Triandis, H.C., Gelfand, M.J. (1998) Converging measurement of horizontal and vertical individualism and collectivism, *Journal of Personality and Social Psychology*, 74 (1), 118–128.

Trompenaars, F., Hampden-Turner, C. (1997) *Riding the Waves of Culture: Understanding Cultural Diversity in Business*, 2nd edn, London: Nicholas Brealey Publishing.

Trompenaars, F., Hampden-Turner, C. (2005) *Riding the Waves of Culture: Understanding Cultural Diversity in Business*, 3rd edn, London: Nicholas Brealey.

Uher, T.E. (1990) The variability of subcontractors' bids, In *Proceedings, CIB90: Building Economics and Construction Management*, University of Technology, Sydney, pp. 576–586.

United Nations (2008) *International Standard Industrial Classification of All Economic Activities (ISIC rev. 4)*, http://unstats.un.org/unsd/cr/registry/regcst.asp?Cl=27 (accessed 17 June 2009).

Venekatte, E. (2002) Open Normen in de Reclame: Een Onderzoek naar de Uitspraken van de Reclame Code Commissie, haar College van Beroep en de Rechter in de Periode van 1982 tot 2000, PhD thesis, University of Twente, Enschede, the Netherlands.

Verma, V.K. (1996) *Human Resource Skills for the Project Manager*, Newton Square, PA: Project Management Institute.

Vise, D.A., Malseed, M. (2005) *The Google Story: Inside the Hottest Business, Media and Technology Success of Our Time*, New York: Bantam Dell, Random House.

Vita Valley, TNO (2010) *Bouwen vanuit de Behoeften van Ouderen*, Delft, the Netherlands: Vita Valley Foundation.

van de Vliert, E. (1998) Conflict and conflict management, In Drenth, P.J.D., Thierry, H., de Wolff, C.J. (eds) *Handbook of Work and Organizational Psychology*, 2nd edn, *Volume 3: Personnel Psychology*, Hove: Psychology Press, pp. 351–376.

Vos, M., de Wit, J.M.A.M., Duivesteijn, A.Th., van der Staaij, C.G., Smulders, H.J.C., van Beek, W.J.J., Pe, M., ten Hoopen, J. (2002) *De Bouw uit de Schaduw: Parlementaire enquête Bouwnijverheid – Eindrapport*, The Hague, the Netherlands: SdU-Uitgevers.

Weeks, J., Gulunic, C. (2003) A theory of the cultural evolution of the firm: the intra-organizational ecology of memes, *Organization Science*, **24** (8), 1309–1352.

Westhead, K., Mortenson, C., Moore, J., Williams Rice, A. (2000) *New Economy: Forget the Web, Make Way for the Grid*, London: Deutsche Bank AG.

Williams, A., Dobson, P., Walters, M. (1989) *Changing Culture: New Organizational Approaches*, London: Institute of Personnel Management.

Williamson, O.E. (1985) *The Economic Institutions of Capitalism*, New York: Free Press.

Womack, J.P., Jones, D.T., Roos, D. (1990) *The Machine that Changed the World*, New York: Rawson Associates.

World Bank (2009) *World Development Report 2009*, http://econ.worldbank.org/WBSITE/EXTERNAL/EXTDEC/EXTRESEARCH/EXTWDRS/EXTWDR2009/0,,contentMDK:21955654~menuPK:4231159~pagePK:64167689~piPK:64167673~theSitePK:4231059,00.html (accessed 27 July 2009).

World Commission on Environment and Development (1987) *Our Common Future (the Brundtland Report)*, Oxford: Oxford University Press.

Yamagishi, T., Yamagishi, M. (1994) Trust and commitment in the United States and Japan, *Motivation and Emotion*, **18** (2), 129–166.

Zein Al-Abideen, H.M. (1998) Concrete practices in the Arabian Peninsula and the Gulf, *Materials and Structures*, **31**, 275–280.

Zhang, S. (2004) An organizational cultural analysis of the effectiveness of Chinese construction enterprises, PhD thesis, The University of Hong Kong.

Index

acculturation–adaptation 47
adhocracy culture 53, 54
Adriaanse, A., Voordijk, H. and Dewulf, G. 160
advanced industrialized countries (AICs) 25
adversarialism 42–3
advice *see* lessons learned
Africa 40, 47
age and culture 58
Agenda 21 (1992) 69
agents of stakeholders 39
Akiner, I. 118
Akiner, I. and Tijhuis, W. 109
Alliance Management International Ltd 27, 64
alliances: culture and 63–8; globalization and 27; strategic purposes of 19
allocation, stakeholders and 38
Alvesson, M. 62
Amason, A.C., Hochwarter, W.A., Thompson, K.R. and Harrison, A.W. 73
Amsterdam, office projects in 169
Anderson Consulting 27, 64
architects: construction industry 37; design in UAE, case study 125, 126–8, 129, 130, 131, 132; inner-city project in Netherlands, case study 87, 88, 89, 90; serial housing in Germany, case study 94; subcontracting in Poland, case study 101–2, 104; tendering in Turkey, case study 110, 112, 114, 116, 117
Arup, Ove 52
Association of South-East Asian Nations (ASEAN) 30
assurance, trust and 66
Au, K.Y. 14, 15, 60

Australia 47, 138
Austria 29

Baan, J. 123, 163
Bachmann, R. 9, 11, 19, 65
Baiden, B.K., Price, A.D.F. and Dainty, A.R.J. 20, 63, 65
Baumol, W.J. 4
Beamish, P.W. 129
Beck, C.H. 98
Beck, W. and Herig, N. 92
behaviour: changes in, uncertainties of 174–5; criminal behaviour 169; critical incidents, behaviour during, developments in construction industry 167–72; culture and 45; global harmonization of regulation, need for 171; individualism within harmonization 174; interconnection and empowerment 171; misbehaviour, risk of 174; multi-cultural behaviour 173; negative information, risk of spread of 171; opportunistic behaviour 102; reputation and 170; scandal 169; social media networks, effect on 170–2; Thomas–Kilmann model of compromise 168–9; Wikileaks™, effect on 172
behaviour: business behaviour, problems and solutions 102–3
Belgium 29, 31
Bernoulli, D. 3
Black, J.S. and Gregersen, H.B. 121
Blanken, A. 118
Boisot, M. and Child, J. 66
Bomers, G.B.J. 158
Bon, R. and Pietroforte, R. 24
van der Boon, V. and van der Marel, G. 169

190 *Index*

Bordieu, P. and Wacquant, L. 10
Boyer, R. 16
Brazil 165
BRE Environmental Assessment Method (BREEAM) 71–2
breadth of culture 46
Bresnan, M. and Marshall, N. 20
Brundtland Report (1987) 69
Building Information Modelling (BIM) 159, 160–1
buildings 2, 6, 9, 12, 15–16, 25, 68, 137; apartment buildings 85; buildings-in-use, work on 4; design of 38; environmental impacts of 72; 'green' performance of 70–1; human welfare, contributions to 5; responsive buildings 61; socio-economic origins of 1
Burns, T. and Stalker, G.M. 56
business alliances 63–5, 66–7; *see also* alliances
business culture 22; behaviour, individualism within harmonization 174; characteristics of 169; construction industry role in 172–7; Culture in Construction (CIB W112) 80, 173; globalization and 173; influence of, need for awareness of 169–70; influence on construction industry 156, 175; market knowledge, personal relationships and 174; misbehaviour, risk of 174; multi-cultural behaviour 173; procedures and processes, differences in interpretation on uses of 173–4, 175; 'think globally, act locally' 174; transparency 173
business development, problems of 141, 142–3, 143–4, 144–5, 146
business partners, selection criteria 110
business relationships, creation of 152
business-to-business (B2B) developments 166

Cameron, K.S. and Quinn, R.E. 53–5, 59, 61
Canada 31
case studies 21, 80–150; approach of authors to 82–3; China, export market development 80, 137–50; construction phases 82; critical incidents, focus on 82; design in United Arab Emirates 80, 121–37; export market development in China 80, 137–50; Germany, serial housing 80, 91–8; inner-city project in Netherlands 80, 83–91; Netherlands, inner-city project 80, 83–91; Poland, subcontracting 80, 98–108; serial housing in Germany 80, 91–8; 'snap shots' of critical moments 150; subcontracting in Poland 80, 98–108; tendering in Turkey 80, 108–21; Turkey, tendering 80, 108–21; United Arab Emirates, design 80, 121–37
categorization: categories of culture 45–6, 56–7; of countries, globalization and 25–6
Chandler, A. 33
Chartered Institute of Building (CIOB) 126
Chen, C.C., Meindl, J.R. and Hunt, R.G. 51
China 3, 4, 21, 28, 29, 30, 31, 40, 51, 59, 165; Chinese Culture Connection 50; economic power in 40; export market development, case study 80, 137–50; networked transactions, system of 66
clan culture 53, 54
Clegg, S.R. 74
client–contractor relationships 158
clients: construction industry 37, 39; design in UAE, case study 125, 126, 129–30, 131–2; export market development in China, case study 141, 144–5, 146; inner-city project in Netherlands, case study 87; serial housing in Germany, case study 94, 95; subcontracting in Poland, case study 102, 103–4, 104, 105; tendering in Turkey, case study 110–11, 112, 113–14, 116, 117–18; turnkey solutions for 108–9
cloud computing 164
Cole, R.J. 70–1
collaborative software tools 163–5
collective programming 44–5
collectivism 142, 145; culture and 50, 51
communication 12, 17, 25, 48, 74, 89, 90, 100, 123–4; channels for 10–11; communication chains 39; communication process 127; cross-border communication 165; design communication 131–2; global communication 27; informal

Index 191

communication 133; lessons learned 151, 152, 154, 155; problems in construction industry 27, 62–3; *see also* information and communication technology (ICT)
compartmentalization 33
competing values and organizational cultures model 53–5
competition: globalization and 165–6; lessons learned 152; within teams 106
complexity of construction industry 3
computer-assisted design (CAD) 160–1
concession arrangements 39
conflict: conflict episode, antecedents and consequences 74–5, 76; conflict management styles 78; conflict (resolution) management 77–8; constructive conflict 74; destructive conflict 74; disputes and 73–8; handling, lessons learned on 154–5; interpersonal conflict 78; prevention, lessons learned on 154–6; resolution, opportunities in 155, 156; sources of 74–5; typology of 73–4
Confucius 142, 145
conglomeration and globalization 23–4, 39–40
consortium 109, 110, 112, 113–14, 116–17, 119
Construct I.T. 161
construction culture 58–9, 61–3; *see also* business culture; culture
construction industry: adversarialism in 42–3; agents of stakeholders 39; allocation, stakeholders and 38; architects 37; behaviour during critical incidents in, developments in 167–72; business culture, role in 172–7; business cultures, influence on 156; client–contractor relationships 158; clients 37, 39; communication problems in 62; compartmentalization 33; complexity of 3; components and practices of, need for improvement in 42; concession arrangements 39; contradictions within processes and procedures 4; corruption 31; corruption, consequences of 33; culture and 16–18, 20, 21; culture and, future for 21–2; demand for (and demand transmission processes) 2; design and, traditional separation of 2; developments in 157–67; division of labour 3; economic cycles 39; economic importance of 1, 4–8; ethics 32; financial power 39; fragmentation of 3, 10, 39, 42, 63; government and 24; governmental investigations 158; identification, stakeholders and 38; industry, functional classifications within 23–4; integration in 3–4; labour-intensiveness of 1; law and codes of conduct 32–3; legislation 39; main contractors 37, 39, 63; market orientation 37; morality 32; organizational development 33–4, 34–5; outputs of 1–2; partnering 63–4; people business 174, 175; performance in 2–3, 4; performance in, objectives and 37; procurement, partnering in 63–4; procurement arrangements, trends and 3–4, 20, 41; professionalism 33; project participants 37; project realization process 35; project size and complexity 37, 40–1; projects as joint ventures 18–20; public–private partnerships (PPPs) 2, 7, 39, 40; quantification, stakeholders and 38; quantity surveyors 37; regulation 37–8; response, stakeholders and 38; risk levels and distributions 3–4; services engineers 37; society, importance for 1–2, 8–11; specialization 3–4, 39; stakeholders 36–7, 38–9; structural engineers 37; structure of industry, dynamic nature of 3; structures and processes 33–4, 34–5; subcontracting 2, 4, 7, 8, 39, 80, 93–5, 98, 153; sustainability issues 158; team building 158; transparency 31; turbulence 36, 63; value accumulation mechanisms 64; visibility 31
construction phases 82
Construction Task Force (1998) 42
constructive conflict 74
consultants: design in UAE, case study 125–6, 127–8, 130, 131, 132–3; inner-city project in Netherlands, case study 87; serial housing in Germany, case study 94; subcontracting in Poland, case study 101–2, 104

Index

Contact–Contract–Conflict (3C) Model 21, 79–80; design in UAE, case study 127, 133–4, 135–6; export market development in China, case study 146–7; framework for investigation of construction processes 79–80, 81; inner-city project in Netherlands, case study 88, 89–90; serial housing in Germany, case study 95–6; subcontracting in Poland, case study 105–6, 106–7; tendering in Turkey, case study 118–19, 119–21; *see also* case studies
contacts, sustenance of 151–3
context of culture 48–9
contra-expertise 102–3
contract-model design 128
contract type, choice of 153
contractors: inner-city project in Netherlands, case study 87; serial housing in Germany, case study 94, 95, 96–7; subcontracting in Poland, case study 100, 104, 105, 107; tendering in Turkey, case study 110, 115
contracts: contractual alliances 65; improvement of, lessons in 153–4
control 66–8
Cooper, R. and Aouad, G. 161
corruption 31; consequences of 33
cost–benefit analysis (CBA) 72–3
cost estimation 128
costs, benefits and sustainability 69, 72–3
criminal behaviour 169
cultural awareness 58
cultural change 53, 59, 60–1
cultural convergence 24–5
cultural differences 57–9
cultural distances 59–60
culture 11–16, 20; acculturation-adaptation 47; adhocracy culture 53, 54; age and 58; alliances 63–8; assurance, trust and 66; attention to, lessons learned 151; behaviour and 45; breadth of 46; business alliances 63–5, 66–7; categories of 45–6, 56–7; clan culture 53, 54; collective programming 44–5; collectivism and 50, 51; comparisons of 47; competing values and organizational cultures model 53–5; conflict, typology of 73–4; conflict and disputes 73–8; conflict episode, antecedents and consequences 74–5, 76; conflict management styles 78; conflict (resolution) management 77–8; construction culture 58–9, 61–3; construction industry and 16–18, 20, 21; constructive conflict 74; consumption of resources and sustainability 70, 73; context of 48–9; contractual alliances 65; control 66–8; costs, benefits and sustainability 69, 72–3; cultural awareness 58; cultural change 53, 59, 60–1; cultural differences 57–9; cultural distances 59–60; definitions and nature of 43–7; Denison's model of organizational culture 55–6; depth of 46; destructive conflict 74; dimensions of, sustainability and 69; dimensions of cultures 48–57; dynamic nature of 44; dysfunctional conflict 74; equity control 67; evaluation issues on sustainability 71–3; femininity and 50; formal and informal organizational structures, dichotomy between 56; forms of sustainability 69–70; functional conflict 74; future for construction industry and 21–2; 'greening', sustainability and 69–71; group aspect of 44; hierarchy culture 54, 55; human behaviour 67–8, 68–9; individualism and 50, 51; indulgence and 50–1; institutional environment and organizational culture, relationship between 67; inter-relations of subcultures 46–7; interpersonal conflict 78; layers of 44–5; long-termism and 50; market culture 53, 54; masculinity and 50; monochronic time and 49; motivation for sustainability 72; national and organizational cultures, relationship between 67; organization charts 56; organizational climate 57; organizational culture 52–7; organizational individualism 51; performance, cultural change and 59; person culture 56; polychronic time and 49; power culture 56; power distance and 50; progression of 46; resolution of conflict and disputes 75–7; resources and alliances 67; restraint and 50–1;

risks in business alliances 66–7; role culture 56; Schein's essence of 45; sequential time and 49–50; sources of conflict 74–5; space and 50; sustainability 68–73; sustainable construction, principles for 70; synchronic time and 49–50; task culture 56; trust 42, 65–6; uncertainty avoidance and 50; value-oriented dimensions of 51–2
culture (contact): design in UAE, case study 135; export market development in China, case study 148; inner-city project in Netherlands, case study 89; serial housing in Germany, case study 96–7; subcontracting in Poland, case study 106–7; tendering in Turkey, case study 119–20
Culture in Construction (CIB W112) 80, 173

Dainty, A.R.J. *et al.* 63
Das, T.K. and Teng, B.-S. 64, 66–7
De Boer, L. 133
decision-making 128–30
Deenstra, T. 169
Denison, D.R. 55, 59; model of organizational culture 55–6
DENMA 142
Denmark 31
deontology 32
depth of culture 46
design: construction industry and, traditional separation of 2; phase in subcontracting in Poland 100; solution in subcontracting in Poland 100–1; tendering in Turkey, case study 113
design in UAE, case study 80, 121–37; architect 125, 126–8, 129, 130, 131, 132; clients 125, 126, 129–30, 131–2; consultant 125–6, 127–8, 130, 131, 132–3; Contact–Contract–Conflict (3C) Model 127, 133–4, 135–6; contract-model 128; cost estimation 128; culture (contact) 135; decision-making 128–30; engineer 125, 131; lessons learned 135–6; plan, stages of 126–7; problems, causes of 131–2; project description 122–5; project organization (contract) 135–6; realism, need for sense of 133;

solar energy, problems of 129; stakeholders 125; stakeholders, behaviour of 125–33; summary 136–7; symbiosis 130–1; technology (conflict) 136
destructive conflict 74
Deutsch, M. 73
developers: inner-city project in Netherlands, case study 86–7, 90; serial housing in Germany, case study 93–4, 95, 96–7; tendering in Turkey, case study 110, 115
developing regions, rise of 165
Dietrich, M. 19
Disney® 161
distributors 141, 144, 145; distribution structure, planning for 142–3
division of labour 3
Du Plessis, C. 158
dysfunctional conflict 74

Earth Summit (1992) 69
Eastern Europe 3
e-commerce 166
economic cycles 39
economic growth, influences on 26
economic importance of construction industry 1, 4–8
Egan, J 158
Egan Report (1998) 42
Eldridge, J. and Crombie, A. 46
energy infrastructure 25
Engineering, Construction and Architectural Management 8
engineers: design in UAE, case study 125, 131; inner-city project in Netherlands, case study 87; serial housing in Germany, case study 94; subcontracting in Poland, case study 101–2, 103–4, 104–5
Enshassi, A. and Burgess, R. 125
enterprise-resource-planning (ERP) 163–4
equity control 67
equity joint-ventures (EJVs) 18–19, 64–5
ethics 32
Europe 29, 98, 166, 175; Central Europe 98, 108; Western Europe 98, 108, 111, 147
European Union (EU) 9, 30, 98, 173, 175
Euwema, M.C. and van Emmerik, I.J.H. 78

export market development in China, case study 80, 137–50; business developer 141, 142–3, 143–4, 144–5, 146; business development, problems of 143–5; clients 141, 144–5, 146; collectivism 142, 145; Contact–Contract–Conflict (3C) Model 146–7; culture (contact) 148; distribution structure, planning for 142–3; distributor 141, 142–3, 144, 145; lessons learned 148–9; problems, causes of 143, 145–6; project description 137–41; project organization (contract) 148–9; relationships 144–5, 146, 147, 148; stakeholders 141; stakeholders, behaviour of 141–6; summary 149–50; technology (conflict) 149
Ezilike, E.I., Perry, J.G. and Hawwash, K. 60

Fd.nl 166
Fellows, R.F. 12, 15, 35, 38, 46, 47, 54, 111
Fellows, R.F. and Seymour, D.E. 80
Fellows, R.F., Liu, A.M.M. and Storey, C. 33, 62
Felstiner, L.F., Abel, R.L. and Sarat, A. 93, 95
femininity 50
Ferrell, O.C. and Weaver, K.M. 33
Fiedler, F.E. 45
financial power 39
Fingar, P. 123, 165
Fisher, R. and Ury, W. 74
Ford, Henry 52
fragmentation 3, 10, 39, 42, 63
France 29
Friedman, M. 4
functional classifications 23–4
functional conflict 74
Furnham, A. 39, 53

Gates, Bill 52
Germany 21, 29; Berlin Wall, collapse of 158; Deutsches Institut für Normung (DIN) 99; serial housing case study 80, 91–8
Giddens, A. 18
Glaister, K.W., Husan, R. and Buckley, P.J. 18–19, 64–5
globalization 9–10, 20–1, 21, 165–7; advanced industrialized countries (AICs) 25; alliances 27; business culture and 173; business-to-business (B2B) developments 166; categorization of countries 25–6; competition and 165–6; conglomeration and 23–4, 39–40; cultural convergence 24–5; developing and less developed countries (LDCs) 25; developing regions, rise of 165; e-commerce 166; economic growth, influences on 26; energy infrastructure and 25; 'glocalization' 26–7; harmonization of regulation, need for 171; human aspects in world of 166–7; income per capita, categorization by 25–6; information infrastructure and 25; of information processing 166; information technology (IT) and integration of markets 28–9; 'insiderization' 26; least developed countries (LDCs) 25; newly industrialized countries (NICs) 25; outsourcing 165; physical development and 25; project-based industries 27; regional 'integration' 29–31; stages of 26–7; structural change and 39–40; transport infrastructure and 25; web-enabled technologies 166
Gomez, C., Kirkman, B.L. and Shapiro, D.L. 51
González-Benito, J. and González-Benito, O. 70
government: construction industry and 24; governmental investigations 158
Gruneberg, S. and Hughes, W. 8, 20
guidance *see* lessons learned

Hagen, J.M. and Choe, S. 65, 66
Hall, E.T. and Hall, M.R. 48–9, 50
Hall, P. 38
Handy, C.B. 56
Harris, L.C. and Ogbonna, E. 60, 61
Harvey-Jones, John 52
Hatch, M.J. 13, 44
Health and Safety Executive (HSE) 5–6
Hellriegel, D., Slocum, J.W. and Woodman, R.W. 67
Henry VI (Shakespeare, W.) 95
Heseltine, Michael 7
hierarchy culture 54, 55
Higgin, G. and Jessop, N. 62
Hillebrandt, P.M. 6, 24
Hinman, L.M. 32

Hofstede, G.H. 9, 26, 32, 44–5, 47, 48, 50, 51, 52, 56, 59, 60, 67, 68, 69, 85, 93, 102, 111, 117, 130, 142
Hofstede, G.H. and Bond, M.H. 142, 145
Hofstede, G.H. and Hofstede, G.J. 130
Hong Kong 29; capitalism in 40
Housing Grants, Construction and Regeneration Act (1996) 18
human behaviour 67–8, 68–9
human capital 11
human globalization 166–7
Hungary 29
Hutton, W. 4, 18, 40

identification, stakeholders and 38
income per capita, categorization by 25–6
India 122, 125–6, 127, 128, 129, 131, 132, 165
individualism 50, 51
indulgence 50–1
information and communication technology (ICT) 21, 158–9, 159–65; Building Information Modelling (BIM) 159, 160–1; cloud computing 164; collaborative software tools 163–5; computer-assisted design (CAD) 160–1; e-commerce 161; enterprise-resource-planning (ERP) 163–4; information infrastructure 25; information translation 161; integration of markets and 28–9; motivation for use of ICT 160–1, 161–2; service-oriented architecture (SOA) 164–5
information processing, globalization of 166
infrastructure 5, 6, 16, 94–5, 137, 158; assessment of 17; energy infrastructure 25; information infrastructure 25; socio-economic origins of 1
inner-city project in Netherlands, case study 80, 83–91; architect 87, 88, 89, 90; city council 87; clients 87; consultant 87; Contact–Contract–Conflict (3C) Model 88, 89–90; contractor 87; culture (contact) 89; developer 86–7, 90; engineer 87; investor 87; land-owner 87; lessons learned 89–90; municipality 87; neighbour groups 85–6, 88, 89, 90; problems, causes of 87–9; project description 83; project organization (contract) 89–90; stakeholders 83–4; stakeholders, behaviour of 84–9; summary 91; technology (conflict) 90
'insiderization' 26
integration: in construction industry 3–4; international integration 26; of markets, IT and 28–9; regional 'integration' 29–31
inter-relations of subcultures 46–7
interconnection and empowerment 171
interior design 4, 36, 87
International Construction 40
International Council for Research and Innovation in Building and Construction (CIB) 126, 173; Task Group TG23 80, 173
International Monetary Fund (IMF) 30
International Organization for Standardization (ISO) 38; ISO 9000 18; ISO 14000 71; ISO 14001 71; ISO 9001/9002 93
internationalization 9–10
interpersonal conflict 78
investors: inner-city project in Netherlands, case study 87; tendering in Turkey, case study 110, 112, 113, 114, 115–16, 116–17
Italy 29

Japan 13, 40; mutual interdependence in 66
Jemison, D.B. and Sitkin, S.B. 60

Kahn, R.L. and Katz, D. 105
Kahneman, D. 3
Kahneman, D. and Lovallo, D. 67
Kahneman, D. and Tversky, A. 3
Katzenbach, J.R. and Smith, D.K. 103
Keynesianism 6–7
Kipling, Rudyard 35
Kirkpatrick, D. 170
Koehn, D. 33
Kogut, B. and Singh, H. 15, 59
Korczynski, M. 67
Kroeber, A.L. and Kluckhohn, C. 13, 43–4
Kyoto Protocol 30, 72

Laan, A. 117
labour-intensiveness 1
Laing, John 52
land-owners 87
landscape architecture 4, 36

Langford, D. and Hughes, W. 126
language (and languages) 16, 58
Latham, M. 18, 42, 158
Lau, E. and Rowlinson, S.M. 66
Laufer, A., Raviv, E. and Stukhart, E. 85
law and codes of conduct 32–3
Lawrence, P.R. and Lorsch, J.W. 3, 63
layers of culture 44–5
Leary, M.R. 32
Lebanon, Republic of 121–2
legislation 39
Lego® 165–6
less developed countries (LDCs) 25
lessons learned: business relationships, creation of 152; clarification of problems 155; communication 151, 152, 154, 155; competition 152; conflict handling 154–5; conflict prevention 155; conflict resolution, opportunities in 155, 156; conflicts, prevention of 154–6; contacts, sustenance of 151–3; contract type, choice of 153; contracts, improvement of 153–4; culture, attention to 151; design in UAE, case study 135–6; export market development in China, case study 148–9; inner-city project in Netherlands, case study 89–90; long-term goals 153–4; long-term relationships 152; mutual agreement 153; negotiation 153, 155; problem solving 155; role separation 153, 154; serial housing in Germany, case study 96–8; stakeholders, behaviour of 152; subcontracting in Poland, case study 106–7; team members, choice of 153; tendering 153; tendering in Turkey, case study 119–21; trust 151–2, 153, 154; working cooperatively 152
Letisko International Airport, Bratislava 162–4
Lewicki, R.J., McAllister, D.J. and Bies, R.J. 66
Li Ka Shing 52
Lipsey, R.G. 11, 19, 25
Liu, A.M.M., Fellows, R. and Tijhuis, W. 80
Lloyd-Jones, T. 122
locationlessness 16
long-termism 50; long-term goals 153–4; long-term relationships 152
Lorentz curve 26

Lucas, C. 3, 14–15

Macau 29
McGregor, D. 57
McKenna, E. 73
McSweeney, B. 51
main contractors 37, 39, 63
management contracting 4
Mao Zedong 142
market culture 53, 54
market knowledge, personal relationships and 174
market orientation 37
masculinity 50
Michel, S. and Beuret, M. 138
Middle East 40, 108, 110, 111, 121, 125, 127, 129, 131
Minkov's World Values Survey 50–1
Mintzberg, H. 142
misbehaviour, risk of 174
monochronic time 49
morality 32
Moran, E.T. and Volkwein, J.F. 57
Morlay, P. and Wallis, R. 174
Mulholland, A. and Earle, N. 123, 163
Mulholland, A., Pyke, J. and Fingar, P. 166
multi-cultural behaviour 173
multi-disciplinary practices 4, 39
municipalities: inner-city project in Netherlands, case study 87; serial housing in Germany, case study 94; subcontracting in Poland, case study 104
Murdoch, J. and Hughes, W. 111
mutual agreement 153

Nahapiet, J. and Ghoshal, S. 11
national and organizational cultures, relationship between 67
Negandhi, A.R. and Reimann, B.C. 130
negative information, risk of spread of 171
negotiation 153, 155
neighbour groups 85–6, 88, 89, 90
Netherlands 21, 29, 31; capitalism in 40; collusion, regulations against in 158; DNR 2005 (*The New Rules 2005: Explanatory Notes on the Legal Relationship Client-Architect, Engineer and Consultant DNR 2005 - Standard Form of Basic Contract;* BNA and ONRI, 2005) 1; inner-city project, case study 80, 83–91

New Zealand 31
Ngowi, A.B., Pienaar, E., Talukhaba, A. and Mbachu, J. 25
Nicolini, D. 63
non-equity joint-ventures (NEJVs) 18–19, 64–5

Odijk, S.W. 125
Office of Government Commerce (OGC) 41
Office of National Statistics (ONS) 5, 6, 24
Ohmae, K. 25, 27
Oliver, C. 67
opportunistic behaviour 102
organization charts 56
Organization for Economic Cooperation and Development (OECD) 30
organizational climate 57
organizational culture 52–7; formal and informal organizational structures, dichotomy between 56; institutional environment and organizational culture, relationship between 67
organizational development 33–4, 34–5
organizational individualism 51
Ouchi, W.G. 57
outsourcing 165
Oxford English Dictionary (OED) 65, 69, 73

Pangakar, N. and Klein, S. 19, 27, 65
partnering 4, 63–4
performance 2–3, 4; cultural change and 59; objectives and 37; subcontracting in Poland, case study 103–4
person culture 56
PIO KMP and NTGroep 162–4
Poland 21; subcontracting case study 80, 98–108
polychronic time 49
Pondy, L.R. 75, 76
power culture 56
power distance 50
Pritchard, J. 33
problem resolution *see* lessons learned
problems, causes of: design in UAE, case study 131–2; export market development in China, case study 143, 145–6; inner-city project in Netherlands, case study 87–9; serial housing in Germany, case study 95; subcontracting in Poland, case study 104, 105; tendering in Turkey, case study 116–18
procurement: arrangements for, trends in 3–4, 20, 41; partnering in 63–4
professionalism 33
project descriptions: design in UAE, case study 122–5; export market development in China, case study 137–41; inner-city project in Netherlands, case study 83; serial housing in Germany, case study 91–2; subcontracting in Poland, case study 98–9; tendering in Turkey, case study 108–9
project management 4, 21, 37
project organization (contract): design in UAE, case study 135–6; export market development in China, case study 148–9; inner-city project in Netherlands, case study 89–90; serial housing in Germany, case study 97; subcontracting in Poland, case study 107; tendering in Turkey, case study 120
projects: as joint ventures 18–20; participants 37; project-based industries 27; project culture, concept of 16, 17–18; realization process 35; size and complexity 37, 40–1
public–private partnerships (PPPs) 2, 7, 39, 40

quantification, stakeholders and 38
quantity surveyors 37

Raiden, A.B., Pye, M. and Cullinane, J. 159
Ray, S. and Chakrabarti, A.K. 130
realism, need for sense of 133
van Rees, R. 161
regional 'integration' 29–31
regulation 37–8
relationships 144–5, 146, 147, 148
Ren, L. 137
reputation, behaviour and 170
research and development (R&D) 56, 61
resources: alliances and 67; consumption of resources, sustainability and 70, 73
response, stakeholders and 38
restraint and culture 50–1

198 Index

Rethinking Construction (Egar, J.) 42
van Riemsdijk, M.J. 86–7
Ring, P.S. and van de Ven, A.H. 117
risk: in business alliances 66–7; levels and distributions 3–4
Robbins, S.P. 74
van der Rohe, Mies 16
role culture 56
role separation 153, 154
Rooke, J., Seymour, D.E. and Fellows, R.F. 62, 63
Rosenthal, R. and Rosnow, R.L. 32
Royal Institute of British Architects (RIBA) 128
Ruddock, L. and Ruddock, S. 159
Russia 31, 165

Samuelson, P.A. and Nordhaus, W.D. 26
Sanders, G.J.E.M. 82, 150, 157
scandal 169
Schein, E.H. 44, 45, 55–6, 68, 82, 150, 157
Schneider, S. and DeMeyer, A. 67
Schneider, W.E. 13, 43, 59
Schwegler, B. 161
Scott, W.R. 9
Scotti, R.A. 169
sequential time 49–50
serial housing in Germany, case study 80, 91–8; architect 94; clients 94, 95; consultant 94; Contact–Contract–Conflict (3C) Model 95–6; contractor 94, 95, 96–7; culture (contact) 96–7; developer 93–4, 95, 96–7; engineer 94; lessons learned 96–8; municipality 94; problems, causes of 95; project description 91–2; project organization (contract) 97; stakeholders 92–3; stakeholders, behaviour of 93–6; summary 98; technology (conflict) 97–8
service-oriented architecture (SOA) 164–5
services engineers 37
Shakespeare, William 95
Shane, S. 67
Shenkar, O. 60
Sherif, M. 78
Sheth, J.N. and Parvatiyar, A. 19, 64, 65
Simon, H.A. 62
Smith, Adam 3
social capital 10–11
social constructivism 18
social institutions 8–9
social media networks, effect on behaviour 170–2
society, importance for construction industry 1–2, 8–11
solar energy, problems of 129
Somalia 31
Soviet Union 3, 14, 29, 102
space and culture 50
Spain 29
specialization 3–4, 39
Der Spiegel 158
St Peter's Basilica, Rome 169
stages of globalization 26–7
stakeholders: construction industry 36–7, 38–9; design in UAE, case study 125; export market development in China, case study 141; inner-city project in Netherlands, case study 83–4; serial housing in Germany, case study 92–3; subcontracting in Poland, case study 99–100; tendering in Turkey, case study 110
stakeholders, behaviour of: design in UAE, case study 125–33; export market development in China, case study 141–6; inner-city project in Netherlands, case study 84–9; lessons learned 152; serial housing in Germany, case study 93–6; subcontracting in Poland, case study 100–5; tendering in Turkey, case study 110–18
Steenhuis, H.J. 137
Steensma, H.K., Marino, L., Weaver, K.M. and Dickson, P.H. 67
Stockdale, M., Campbell, C., Naslin, S. and Wong, A. 166
structural change and globalization 39–40
structural engineers 37
structure of industry, dynamic nature of 3
structures and processes 33–4, 34–5
subcontracting 2, 4, 7, 8, 39, 80, 93–5, 98, 153
subcontracting in Poland, case study 80, 98–108; architect 101–2, 104; business behaviour, problems and solutions 102–3; clients 102, 103–4, 104, 105; competition within teams 106; consultant 101–2, 104; Contact–Contract–Conflict (3C) Model 105–6, 106–7; contra-expertise

102–3; contractor 100, 104, 105, 107; culture (contact) 106–7; design phase 100; design solution 100–1; engineer 101–2, 103–4, 104–5; lessons learned 106–7; municipality 104; opportunistic behaviour 102; performance 103–4; problems, causes of 104, 105; project description 98–9; project organization (contract) 107; stakeholders 99–100; stakeholders, behaviour of 100–5; summary 107–8; technology (conflict) 107
Sugden, J.D. 24
Sun Tzu 142
sustainability 2; culture 68–73; dimensions of culture, sustainability and 69; evaluation issues on 71–3; forms of 69–70; 'greening' and 69–71; issues of 158; motivation for 72; sustainable construction, principles for 70
Sweden 31
symbiosis 130–1
synchronic time 49–50
systems theory 13–14

Tagiuri, R. and Litwin, G.H. 57
task culture 56
Tavistock Institute of Human Relations 10–11, 20, 43
Taylor, F.W. 3
team building 153, 158
technology (conflict): design in UAE, case study 136; export market development in China, case study 149; inner-city project in Netherlands, case study 90; serial housing in Germany, case study 97–8; subcontracting in Poland, case study 107; tendering in Turkey, case study 120–1
Teece, D.J. 67
Tempelmans-Plat, H. 158
tendering in Turkey, case study 80, 108–21; architect 110, 112, 114, 116, 117; business partners, selection criteria 110; clients 110–11, 112, 113–14, 116, 117–18; clients, turnkey solutions for 108–9; consortium 109, 110, 112, 113–14, 116–17, 119; Contact–Contract–Conflict (3C) Model 118–19, 119–21; contractor 110, 115; culture (contact) 119–20; design 113; developer 110, 115; investor 110, 112, 113, 114, 115–16, 116–17; lessons learned 119–21; location, search for 113; problems, causes of 116–18; project description 108–9; project organization (contract) 120; stakeholders 110; stakeholders, behaviour of 110–18; summary 121; technology (conflict) 120–1; tender process, negotiations within 110–14; tendering procedure 111
Thomas, K.W. 77, 78, 168
Thomas–Kilmann model of compromise 168–9
Tijhuis, W. 79–80, 81, 93, 95, 111, 127, 129, 138, 146, 159, 169, 173
Tijhuis, W. and Fellows, R.F. 82–3
Tijhuis, W. and Lousberg, L. 133
Toyota 13, 42
transparency: business culture 173; construction industry 31
Transparency International 31, 33
transport infrastructure 25
Triandis, H.C. 51
Triandis, H.C. and Gelfand, M.J. 51
Trompenaars, F. and Hampden-Turner, C. 32, 46, 49, 51–2, 69
trust: culture 42, 65–6; lessons learned 151–2, 153, 154
turbulence 36, 63
Turkey 21, 29; tendering case study 80, 108–21

Uher, T.E. 8
uncertainty 19, 36, 47, 53, 65; changes in behaviour, uncertainties of 174–5; uncertainty avoidance (UA) 50, 56, 60, 93; uncertainty reduction 85, 102, 111
United Arab Emirates (UAE) 21, 122, 128–9; design case study 80, 121–37
United Kingdom 29, 31, 58; arbitration in 76–7; capitalism in 40; construction industry classification in 23–4; economy of 5–6; Egan Report (1998) 42; Environment, Transport and Regions, Dept for (DETR) 70; Health and Safety Executive (HSE) 5–6; Housing Grants, Construction and Regeneration Act (1996) 18; Latham's review of construction in 42; Office of Government Commerce (OGC) 41; Office of National Statistics (ONS) 5, 6, 24;

public housing provision 7; Royal Institute of British Architects (RIBA) 128; team-building approaches in 158
United Nations (UN) 23–4, 29–30; Rio de Janeiro Conference on Environment and Development (1992) 69
United States 13, 29, 31, 42, 58; capitalism in 40
USSR *see* Soviet Union

value accumulation mechanisms 64
value-orientation 51–2
Venekatte, E. 124
Verma, V.K. 168
Vise, D.A. and Malseed, M. 170
visibility 31
Vita Valley and TNO 158
van de Vliert, E. 73, 75, 76
Vos, M. *et al.* 85, 158

Weeks, J. and Gulunic, C. 60
Westhead, K., Mortenson, C., Moore, J. and Williams Rice, A. 161
Wikileaks™ 170; effect on behaviour 172
Williams, A., Dobson, P. and Walters, M. 56
Williamson, O.E. 19, 65
Womack, J.P., Jones, D.T. and Roos, D. 13, 41, 66
World Bank 25–6, 30
World Commission on Environment and Development (1987) 69
World Economic Forum 30
World Trade Organization (WTO) 30
WT/Projects 83, 84, 85, 86

Yamagishi, T. and Yamagishi, M. 66

Zein Al-Abideen, H.M. 125
Zhang, S. 59